Understanding Just Sustainabilities from Within

'*Understanding Just Sustainabilities from Within* provides a thoroughly engaging account of the practice of food justice and non-profit community organizing. Godfrey's activism and intellection engage thought, heart, emotion, intuition, and spirit, enabling her creation of both compelling story and important theory. This book is perfect for my graduate and undergraduate classes on Environmental Justice and Ecofeminism.'

Jane Caputi, *Professor, Center for Women, Gender and Sexuality Studies, Communication and Multimedia, Florida Atlantic University*

Written by the co-founder and former board president of a non-profit shared-use commercial kitchen, *Understanding Just Sustainabilities from Within* presents an intersectional analysis of CLiCK (Commercially Licensed Co-operative Kitchen), in order to explore what just sustainabilities can look and feel like from *within* and *without*.

Through a unique combination of autoethnography, participant observation, surveys, and secondary research, this book offers insights into CLiCK's micro and macro successes, failures, and unknowns in relation to its attempt to put the concept of just sustainabilities into daily practice and praxis. Developing its practical analyses from a theoretical basis, this book does not focus on definitive answers, recognizing instead that the closest we can get to understanding just sustainabilities in praxis is through long-term collective struggle and ultimately love.

Researchers and educators who are interested in linking theory with practice, especially in relation to just sustainabilities and intersectionality, will appreciate the theoretical grounding, making it desirable for multiple social science classes. Additionally, those involved with the social justice, food justice, and just sustainabilities movements will benefit from the book's insights into best practices to address issues of social inequalities on the micro level, while also offering the benefits of a macro intersectional analysis.

Phoebe Godfrey is an Associate Professor in Residence of Sociology at the University of Connecticut, USA.

Routledge Equity, Justice and the Sustainable City

Series editors: Julian Agyeman and Stephen Zavestoski

This series positions equity and justice as central elements of the transition toward sustainable cities. The series introduces critical perspectives and new approaches to the practice and theory of urban planning and policy that ask how the world's cities can become 'greener' while becoming more fair, equitable and just.

The *Routledge Equity Justice and the Sustainable City* series addresses sustainable city trends in the global North and South and investigates them for their potential to ensure a transition to urban sustainability that is equitable and just for all. These trends include municipal climate action plans; resource scarcity as tipping points into a vortex of urban dysfunction; inclusive urbanization; "complete streets" as a tool for realizing more "livable cities"; the use of information and analytics toward the creation of "smart cities".

The series welcomes submissions for high-level cutting edge research books that push thinking about sustainability, cities, justice and equity in new directions by challenging current conceptualizations and developing new ones. The series offers theoretical, methodological, and empirical advances that can be used by professionals and as supplementary reading in courses in urban geography, urban sociology, urban policy, environment and sustainability, development studies, planning, and a wide range of academic disciplines.

Disruptive Transport
Driverless Cars, Transport Innovation and the Sustainable City of Tomorrow
Edited by William Riggs

Understanding Urban Cycling
Exploring the Relationship Between Mobility, Sustainability and Capital
Justin Spinney

Understanding Just Sustainabilities from Within
A Case Study of a Shared-Use Commercial Kitchen in Connecticut
Phoebe Godfrey

For more information about this series, please visit: www.routledge.com/
Routledge-Equity-Justice-and-the-Sustainable-City-series/book-series/EJSC

Understanding Just Sustainabilities from Within

A Case Study of a Shared-Use Commercial Kitchen in Connecticut

Phoebe Godfrey

Routledge
Taylor & Francis Group

LONDON AND NEW YORK

First published 2021
by Routledge
2 Park Square, Milton Park, Abingdon, Oxon OX14 4RN

and by Routledge
605 Third Avenue, New York, NY 10158

Routledge is an imprint of the Taylor & Francis Group, an informa business

British Library Cataloguing-in-Publication Data
A catalogue record for this book is available from the British Library

Library of Congress Cataloging-in-Publication Data
A catalog record has been requested for this book

ISBN: 978-1-138-61501-4 (hbk)
ISBN: 978-1-032-01356-5 (pbk)
ISBN: 978-0-429-46344-0 (ebk)

Typeset in Goudy
by codeMantra

On the personal level...

This book is dedicated to my wife and partner, co-founder of CLiCK, without whom none of this work would have begun, nor could have happened. Love is one of the only forces in the universe that increases the more you share it and certainly our love has increased from all that we have shared with others and all that the others have shared with us. This has been a collective and collaborative project from the beginning, and I am deeply grateful to one and all who have in big and small ways made it possible.

I'd also like to dedicate my commitment to social justice to my mother, who as a trained occupational therapist focusing on people in institutions with mental health problems taught me early on that what matters most is how you treat each and every individual regardless of what personal or social challenges ail them. Central to her work was helping people with mental illness return to community living in supported and dignified ways. Later on, at the end of her career, she created and ran a senior citizen center in Princeton, New Jersey. This center was a model for me in the creation of CLiCK, in that it was a place full of energy, art, and community events – a place where she made everyone feel that they belonged and they mattered. As a result, it was a *Real* place.

Finally, I'd like to remember a dear friend – John D. Tibbetts – who passed away too suddenly and too young while I was writing this book. In response to having read my self-published illustrated story A *North Pole Tale* (2018)[1] about Santa discovering climate change, he wrote, not long before he died, "...write more books...".

On the political level…

I'd like to dedicate this book to the social/environmental justice activists both in the USA and around the world – in particular those who are Indigenous – many for whom such work is *individually* life threatening, even as the for-profit destruction and social oppressions they stand up against ultimately threaten our *collective* lives on earth.

May we not lose *heart*, for the shared *dreams* of social–environmental justice will prevail: and the Earth will endure with *or* without us.

Note

1 Godfrey, P. and FeiFei. 2018. *A north pole tale*. Mansfield, CT: Ellipsis Publishing.

Contents

PART III
Thresholds of successes, failures, and
unknowns – explorations in *praxis*   97

Figures

Foreword

"It had been built in the 1970s and was energetically stale as were the overwhelming smells of beer and cigarettes still emanating from the 'old men's bar'…The commercial kitchen was dirty and decrepit, and the banquet room – complete with bingo board and cheap chandeliers – was depressing. Yet the space clearly had potential…." While most of us would see such a building – a vacant Knights of Columbus hall in Willimantic, Connecticut – and beat a hasty retreat, it became for Phoebe Godfrey, a self-described "social justice activist and social justice academic", *such stuff as dreams are made on*. The "stuff" in this case would come to be known by the acronym for "Commercially Licensed Co-operative Kitchen" – CLiCK. The "dreams" – threatening, like bold hopes and visions often do, to become nightmares – is the insightful and constantly reflective tale of this book and the journey of its tenacious author and her wife, Tina.

Many of us, including this writer, will throw our lives into creating new social enterprises that we fervently believe are some part of the answer to the Earth's social and economic injustices. But few of us will also muster the guts to stand back, look analytically at what we've done, and reveal with a brutal, even confessional kind of honesty what we've learned (the hard way, because there's no other way). I would acquire some of those lessons myself while leading the non-profit Hartford Food System for 25 years on its journey to bring a just and sustainable food system to that Connecticut city. It was during that period that I also came to know Willimantic, the place where the CLiCK story would later unfold in response to the community's deindustrialization and a rising poverty rate that would leave it with the inglorious distinction as Connecticut's poorest city.

As we know now, food insecurity as well as diet-related illnesses follow hot on the heels of poverty, and both wove themselves into Willimantic's unraveling fabric. The all-too-typical response to these problems, which are symptoms of the failures of conventional capitalist structures, has been charity, earnestly offered but ultimately landing with as much impact as a wiffle ball. Where were the jobs? Where were the new businesses? Where was the robust democracy that would steer the community to a just, prosperous, and sustainable future for all?

Out of the bountiful imaginations of CLiCK's co-founders, some ten years ago, sprung the notion that a shared-use kitchen and culinary business incubator could harness the power of food to revive an ailing place. Armed with a mission statement to "Grow, Cook, and Share", and an unwavering commitment to empower lower income community members, CLiCK would garden, instruct, and train in various aspects of food sciences and nurture the capacity of others to create commercially viable food businesses. Its purpose and spirit were best expressed by one of CLiCK's Latino board members: "We need to let our community brothers and sisters know that we are there for them, that we are there to help them achieve their goals and take off as independent and sustainable business people regardless of their background, native language, race, or gender".

Those of us who have also dared to imagine a vision like the one CLiCK offers know that the struggle to attain it takes us down a long, twisting road, cratered with pitfalls and cross-hatched with speed bumps. Fortunately, Dr. Godfrey, who is a sociology professor at the University of Connecticut, brings us a rich trove of hard-won lessons accessed from multiple sources. Combining a healthy dose of social theory with "hands-on", CLiCK-inspired practice allows us to not only answer the question of how to do it, but more importantly, why should it be done that way, especially if racial, gender, and class equity is both our means and end. But what makes this comprehensive case study stand out more than others I've witnessed is the personal journey that Dr. Godfrey takes us on. As a co-founder and member of CLiCK's board of directors, a social systems scholar, and Willimantic resident, she and her wife, Tina, are heavily invested not only in the organization's outcomes but the need to transparently share their experiences – the good, the bad, and the ugly. Her work in its totality brings to mind C. Wright Mills' instructive methodological proclamation in *The Sociological Imagination*, "that social observation requires high skill and acute sensibility; that discovery often occurs precisely when an imaginative mind sets itself down in the middle of social realities".

As the food movement diversifies and expands, especially in response to COVID-19 laying bare the failures and flaws of our dominant, industrialized food system, projects like CLiCK will become more plentiful out of necessity. After all, when food-based social enterprises, inspired and led by social entrepreneurs, can provide new outlets for locally processed and grown food while making an impact on local economies, they should be encouraged to blossom hither and yon. But it's not just the management, marketing, and financing of such enterprises that matter (and believe me, they do), it's the people who are most affected by the food system's injustices who must have a seat at the table. It is in that struggle for social justice – to make it manifest in all that we do – where the CLiCK story matters most.

After weathering every imaginable obstacle that a non-profit organization can endure, it is no small achievement to be able to say, as Dr. Godfrey does, that CLiCK's greatest success is "We're still open". But success comes in many forms, such as the Mexican immigrant woman who created a thriving Mexican food catering business out of CLiCK's kitchen and now employs several workers. Success also emerges when you fight tooth and nail to make your enterprise inclusive, community-driven, and respectful of the multi-faceted ways that the human race reveals itself, when you could just as easily have found escape in your white privilege. And there is the success that transcends all else when you learn from your failures, struggle to find answers, and share with the world what you've learned. That just may be this book's most important message.

Mark Winne
November 3, 2020

Preface

CLiCK's Interfaith Land Blessing

This is from CLiCK's Interfaith Land Blessing held Sunday, December 7, 2014, to ensure the *spirit* of CLiCK began and remained one of inclusion and connection through food and ultimately the land. The sentiment expressed here still applies.

Join us in sharing the vision of CLiCK while participating in a blessing of the space – to bring your best hopes and wishes, to show reverence for the land, and to gather all the energies of those who bring their blessings, who are lending their hearts, souls, hands, and time to and for this project.

Please consider bringing an object of relevance and importance to you, as a symbol of your personal hopes and your community's hopes for CLiCK.

Each symbol will find a home on the property and serve as a reminder of our collective care and gratitude for the land.

Some examples of such an object are – though not limited to – a rock, feather, crystal, homemade object, a symbol of religious devotion…

Our ceremony will be led by "Painted Turtle", Mohegan Elder and Pipe Carrier
Hope to see you!

My involvement had more to do with offering what I know, honoring the Earth through ceremony.
Middle-class male Mohegan Elder

Acknowledgments

In relation to CLiCK

As stated in the dedication, 'this has been a collective and collaborative project from the beginning, and I am deeply grateful to one and all who have in big and small ways made it possible'. However, there are a few key individuals I would like to acknowledge and thank in relation to the CLiCK project, for without their specific energetic and/or financial support CLiCK would not have come to be.

The first is Alta Lash, founder and Executive Director of United Connecticut Action for Neighborhoods (UCAN) and teacher of grassroots organizing at the University of Connecticut School of Social Work, who worked tirelessly with Tina and I in the early days and helped us get connected with our first financial support.

The second is Peter Fish, our neighbor and friend, who upon being invited to inspect a possible building for CLiCK spontaneously offered from his heart to lend CLiCK funds to purchase the building and then served on the Board for the following five years.

The third is Eileen Ossen of the Jeffrey P. Ossen Foundation, whose heartfelt and unwavering commitment to CLiCK has helped to ensure our ongoing success and growth.

In relation to this Book

I would like to thank my colleague and friend Bandana Purkayastha at the University of Connecticut for her ongoing support of my work and of me as an individual.

I'd also like to thank Julian Agyeman (and his co-editors) at Tufts University for previously publishing two chapters I wrote on CLiCK, both of which helped me develop my ideas for this larger full-length project focusing on his seminal work in *just sustainabilities*.

I'd also like to thank my graduate assistant Mary Buchanan, with whom I have also recently co-edited another book focusing on just sustainabilities and whose intellectual depth and editing skills have been invaluable.

I'd also like to thank Jane Caputi at Florida Atlantic University for her helpful feedback and for her intellectual and political comradery.

I'd also like to thank Hedley Freake for his ongoing commitment to CLiCK as a Board member and for his thoughtful feedback.

Finally, I'd like to thank Mark Winne for being willing to write a Foreword and for the ongoing work he has done toward helping to achieve and promote food justice in words and in deeds.

Introduction

Understanding a socially constructed world through my positionality

What is life? A madness. What is life? An illusion, a shadow, a story. And the greatest good is little enough; for all life is a dream, and dreams themselves are only dreams.

¿Qué es la vida? Un frenesí. ¿Qué es la vida? Una ilusión, una sombra, una ficción, y el mayor bien es pequeño.¡Que toda la vida es sueño, y los sueños, sueños son!

Pedro Calderon de la Barca

Nothing that is worth doing can be achieved in our lifetime; therefore, we must be saved by hope.

Nothing which is true or beautiful or good makes complete sense in any immediate context of history; therefore, we must be saved by faith.

Nothing we do, however virtuous, can be accomplished alone; therefore, we must be saved by love.

No virtuous act is quite as virtuous from the standpoint of our friend or foe as it is from our standpoint. Therefore, we must be saved by the final form of love which is forgiveness.

Reinhold Niebuhr

My story

This book is my story, parts of which have been told in various forms before (Godfrey 2017; Godfrey and Torres 2020; Godfrey and Freake 2016), but this is the most current and complete version. This is my "*intellectual activism*" (Collins 2013), 'my madness', my story of an 11-year ongoing project that my wife Tina and I embarked upon, which in and of itself has been an exercise in, as Niebuhr states, "hope", "faith", "love", and "forgiveness" (1952, p. 63). If all goes well, this story will not be finished in our lifetimes, but will continue to evolve along with the needs to 'grow, cook, and share' that are an integral part of our physical survival as humans, as well as the survival of our cultures, economies, and identities. Such needs – and the theme of survival – have become increasingly salient in these uncertain times of a global pandemic, as well as increasing global inequality (UN

News 2020), environmental destruction (Greenfield 2020), and climate change (Figueres and Rivett-Carnac 2020).

Our project has been to co-create and continue growing the 501c3 non-profit known as CLiCK (Commercially Licensed Co-operative Kitchen) in the Town of Windham, in Windham County, in the northeastern corner of Connecticut that is often overlooked by state funding, development, and attention. The center of Windham is called Willimantic (a Pequot-Mohegan name, which possibly means 'place of the swift running water'[1]), where Tina and I also happen to live and which was once the cotton thread capital of New England. Now, it is de-industrialized and ranked as the poorest town in CT, with a poverty rate of 29.9% (Stebbins 2019), more than twice the state average. In this structurally challenged economic context, CLiCK's mission is to:

> **Grow, Cook, Share**: To grow the vitality of our local economy and community by offering shared-use commercial kitchens to farmers and culinary entrepreneurs seeking to create food-based businesses; and to improve the health of our local community by teaching gardening, culinary arts, nutrition, and other food-related classes.[2]

Combining economic development and health–nutrition education makes CLiCK unique (Godfrey and Freake 2016) and very ambitious! This mission faces ongoing challenges and has yet to be fully realized, but we are certainly closer to doing so than when the idea first captured our then *naïve* imaginations. In addition, although at the time we wrote this mission we did not explicitly state that we sought to achieve it through the daily practice of food/social justice,

CLiCK entrance and sunflowers

we tacitly understood that an emphasis on *justice* would be part of both the means and the ends, although this understanding deepened over the years, as will be discussed.

Over the course of CLiCK's history, many people have gotten involved at different stages, but only Tina and I have stuck with it since the beginning (first as founders, then as board members, and now as community advocates). Looking back, I've asked, why and how have we been able to keep this idea going? Granted, it has been an incredibly difficult project from the beginning through today, but our collective answer, of late, is that at the time we believed (and still believe) that it was a 'good idea'. And it *was* a 'good idea'! So 'good' that CLiCK has in fact become an even *better* material reality – but only because we stuck with the 'goodness' of the idea and through our time and effort engaged in a form of collective 'secular transubstantiation', taking the essence of our ideas and making them flesh. This 'flesh' now consists of a 5,600 sq ft building on 2.5 acres that hosts two well-equipped shared-use commercial kitchens, two large walk-in coolers, storage space for small food businesses, a teaching kitchen, five part-time staff, a 30-tree orchard (until the beavers recently cut down 11 trees – see postscript), honey bees, a community garden, and many other daily activities and operations on the part of our staff, members/kitchen users, volunteers, and students, all of which this book seeks to reflect upon and explore. But I am getting ahead of myself, for I have yet to socially situate myself within and akin to my own story.

Earlier, I emphasized that this is *my* story, as everything in this book is written from my perspective and I make no claims to being 'objective', nor would I claim that I alone have created CLiCK. In fact, I would argue that once we recognize that all of our experiences are interpreted through our cultural and social positions, we come closer to being able to glean authentic personal and social 'truths'. In other words, the more we reveal and analyze our social positions, the more we are able to piece together the internal and external 'realities' through which we all construct and have constructed our identity-perceptions. I am hyphenating 'identity-perceptions' as I see them as inseparable; who we are, culturally and socially, is inseparable from how we interpret the world, including ourselves, and from how that outside world/others interpret us. This notion that there is no pure objectivity is not the same as saying there are no social truths, empirical facts, or culturally agreed upon ways of interpreting the material world. Rather, what I seek to identify within the concept of perception, both on the individual and collective levels, is the insight by the sociologist Dorothy Smith, who recognized that the only way of "knowing a socially constructed world is from within" (Smith 1990), meaning that knowledge begins with the perspective of the knower. Smith, as a white female feminist sociologist, came to recognize how much gender shaped perceptions of self and others (although, as a middle-class white woman with racial and class privilege, she did not focus at first on these other identities), as well as vice versa; she and other feminist theorists referred to this critical recognition as 'standpoint theory'. In my case, I seek to push beyond Smith's adage, to focus not just on 'knowing' but on 'understanding', recognizing that my experiences in co-creating CLiCK are not just intellectual. Indeed, my experiences and those of others go deeper than the conceptual level, down into our hearts and bodies, to a level of emotional identification and radical empathy "in service to social justice", in line with Patricia Hill Collins' (2013, p. ix) notion of "intellectual activism". By this I mean that I have intentionally worked to challenge our socially ascribed divisions along the lines of class, race, gender, sexuality, etc., which lead to inequality and oppression, as well as between theory and praxis. As a result, I am not just in the role of the academic engaging in an ethnographic study of CLiCK, but rather I have been an 'insider' who is also an 'outsider'. I am engaging in both an autoethnography and a collaborative ethnography (Lassiter 2004), by consulting with Tina and many others who have played parts over the years in the *transubstantial* formation of CLiCK.

My goal here is to apply a "reflexive" lens (DuPuis et al. 2011; Godfrey 2017) to critically analyze my individual and our collective memory in order to discern whether or not *we* (as opposed to merely *I*) have succeeded in creating, experiencing, and sustaining on a small-scale local level what Julian Agyeman and others (2003, 2013) have referred to as 'just sustainabilities' (JS) as defined by its four principles (explored in the next chapter). If yes, then I seek to propose answers as to how have we done it, how do we know we have done it, have we done it for everyone and in all aspects of our operations, and what valuable lessons have we learned from the process of doing so? Contrastingly, if we have not succeeded in doing so, if not for everyone, or if not all the time, then again, I seek to propose answers as to how not, why not, and how can we engage in different practices to ameliorate these seemingly unjust/unsustainable outcomes? These are not mere intellectual engagements, nor completed achievements that can be quantitatively measured. Rather, sustainability and justice are somatic, emotionally vibrant qualitative states that must continually be experienced, felt, and collectively lived if they are to have any genuine, perceivable meaning (brown 2017). This brings the issue of scale into play, as all experiences and spaces are layered, involving, as David Pellow (2016) remarks in relation to Environmental Justice (EJ), "…multiple scales, from the cellular and bodily level to the global level and back" (p. 4). Therefore, as co-founder and ex-board president, I am very much an 'insider' deeply aware of the complex interpersonal dynamics at work at CLiCK, enabling me to engage in a reflexive critical micro analysis. This micro analysis will include not only examining key choices made over the years but also how they were made and by whom and whether the results were beneficial and/or detrimental to CLiCK's mission and to the individuals involved. However, I am not so 'inside' or entangled (Barad 2007) that I can't also engage in an outsider's more critical perspective, especially given that I am also a sociologist trained in the critical analysis of society, social institutions, groups, and individuals. Therefore, I am also an 'outsider' who is equally aware of social structures and ideologies and can engage in a macro analysis of society including our food systems. Moreover, I am interested in the complex ways and means in which the micro and macro social worlds intersect in overt and covert ways, shaping and reshaping the threads of our personal and political landscapes.

Bringing together an 'insider's' micro analysis and an 'outsider's' macro analysis creates what the sociologist C. W. Mills (1959) referred to as the 'sociological imagination', the ability to see 'personal troubles' within the context of 'social issues' and recognize them as interconnected and inseparable. Additionally, early in my career, I was influenced by the sociologist Michael Burawoy, who called upon his fellow sociologists, as the then-president of the American Sociological Association, to become "public sociologists" so that we might stay true to our "... original passion for social justice, economic equality, human rights, sustainable environment, political freedom or simply a better world, that drew so many of us to sociology" (2005, p. 5). My insider/outsider roles at CLiCK are how I have chosen to respond to this call, to be a public sociologist, to exercise my 'sociological imagination', and to ultimately attempt to stay true to my 'original' and ongoing 'passions' as Burawoy identified. Furthermore, the particular ways and means I have chosen to do so in relation to the food system are also inseparable from my positionality (Acevedo et al. 2015; Alkon and Agyeman 2011; Milner 2007; Rose 1997) and my personal and political opposition to an industrial food system that, like the rest of our society, is dominated by the tenets of capitalism, racism, sexism, and an unequivocal disregard for the sacredness of all life on earth.

Positionality (Acevedo et al. 2015; Alkon and Agyeman 2011; Milner 2007; Rose 1997) recognizes that personal and collective knowledge is socially shaped by the intersections (Collins 2009; Crenshaw 1989, 1991) of our cultural and social identities, such as race, gender, social class, nationality, abilities, and other identity markers. Furthermore, all of these markers are relational and contextual within an individual and between individuals and communities, rather than being essential fixed qualities. This means that my positionality as a white, culturally British/European (British parents; I grew up in Europe from 1970 to 1978), middle-class, progressive, academic female who is married to a woman, with whom I eclectically practice pagan-esque spirituality, has shaped and continues to contextually shape how I experience myself/myself in society and how others experience me/us. Tina shares some of these traits, but differs from me socially in that she was born and has always lived in Connecticut, is from the working-class, was a single-mother of a now grown son, and is non-college educated. Both of our identities have shaped and continue to shape the ways and means in which CLiCK has emerged and therefore they are evolving aspects of our/my story. In addition to shaping my positionality, my identities have intersected with the positionalities of everyone else who has been involved with CLiCK and the communities from which CLiCK has emerged, helping to shape the type of organization into which CLiCK has evolved. In fact, as Rose (1997) eloquently argues, positionality seeks to achieve a form of "transparent reflexivity", wherein "...the power of the academic to produce knowledge can be situated", while still recognizing that there also exists,

> ...a much more fragmented space, webbed across gaps in understandings, saturated with power, but also, paradoxically, with uncertainty: a fragile and

fluid net of connections and gulfs. Seen from this perspective, the research process is dangerous. It demands vigilance, a careful consideration of the research process: another kind of reflexivity, in fact, but one which can acknowledge that it may not be adequate since the risks of research are impossible to know (p. 317).

In short, as Rose goes on to articulate in her conclusion, "We cannot know everything...". More specifically, I cannot know everything about the story of CLiCK, but in acknowledging this, I may ultimately be able to recognize the "…. absences and fallibilities…" of my story, while nonetheless "…recognizing that the significance of this does not rest entirely in our own hands" (p. 317).

As a public sociologist committed to continually engaging my sociological imagination in order to theorize putting social justice into *praxis*, my view of society continually evaluates social ideologies, interactions, and institutions through an intersectional lens. Intersectionality (Collins 2009; Crenshaw 1989, 1991) is the theoretical framework shaping positionality; it recognizes that all aspects of our social identities (including our relationships to the natural world [Godfrey and Torres 2016a], explored more in Chapter 2) are interconnected and cannot be analyzed nor understood separately from each other. As such, what Tina and I (and later, others) have done in working to co-create CLiCK has been an attempt to consciously bring an intersectional analysis to our social justice practice, while recognizing that our daily practice and ultimately our *praxis* are not separate from our identities of being white, nor from being women, nor from being a married couple, nor from practicing pagan-esque spirituality, nor from being progressives, nor from being, in my case, a public sociologist, nor from the geographic location in which we live. This is not to claim that creating CLiCK was our 'destiny' or that we have fully achieved our goals, but to recognize upfront, from the perspective of my/our positionalities, the social patterns that have influenced my/our actions and those of others who are part of this story, as told by me, hence part of *my* story.

In looking at social patterns, it becomes apparent that people within given social contexts are less 'individual' than we like to believe; we are all continually being socially shaped and therefore we behave within and through prescribed social expectations, opportunities, and limits. The eminent French sociologist Emil Durkheim was one of the first to theorize that individuals behave according to identifiable patterns based on such social variables as sex-gender, social class, race, religion, etc. Likewise, when it comes to who engages in the various aspects of the Alternative Food Movement (AFM), discernable patterns emerge based on intersecting social identities and their relations of privilege and/or oppression. Simply put, the AFM has been largely a white middle-class movement (e.g., Alkon and Agyeman 2011; Bradley and Herrera 2016; Guthman 2007, 2008; Slocum 2007), inspired in part by popular and socially influential white food writers beginning with Francis Moore Lappé (*Diet for a Small Planet* [1971]), and gaining more traction in terms of larger environmental issues from the works of Wendell Berry (*The Unsettling of America* [1977] and then his more food-specific book *Bringing it to the Table* [2009]), as well as Barry Commoner (*The Closing Circle* [1971]), and more recently, the works of Eric Schlosser (*Fast Food Nation* [2001]), Marion Nestle (*What To Eat* [2006]), and Michael Pollan (*Omnivore's Dilemma* [2006]), all of whom brought attention to negative impacts of the industrial food system, both on people's health (often white people's health in particular) and on the health of the planet (most notably, the areas where white farmers/people live). Members of the predominantly white middle-class public who were influenced by these works sought 'alternative' sources of food such as organic, locally grown, and specialty foods sold at farmers' markets, co-operative stores, or more recently in upscale grocery stores like Whole Foods. As such, the AFM tends to focus on the individual consumer (as in those who are white middle-class) and their food choices and practices, as opposed to seeking more systemic, hence political and thereby more challenging collaborative solutions. In my case, being white and middle-class and a reader of all the aforementioned authors, as well as being economically able to purchase 'alternative foods', I fit well into this AFM stereotype.

However, as a social justice activist and as a sociologist, I was also exposed to other well-known theorists, writers, and activists either from the Global South or who identify as Black Indigenous People of Color (BIPOC), such as Raj Patel (*Stuffed and Starved* [2012]), Vandana Shiva (*Stolen Harvest* [2000]), and more recently, Leah Penniman (*Farming While Black* [2018]), as well as many others who are rightfully more focused on activism/farming than writing (such as Karen Washington[3] and Malik Yakini,[4] to name just two). I am therefore very aware of how and why the AFM has been rightfully critiqued by the Food Justice Movement (FJM) for failing to make direct connections between the negative health and environmental repercussions of the industrial food system and structural racial/class/gender inequalities and injustices within the entire society, including the food system. Unsurprisingly, the FJM has been mostly spearheaded by BIPOC, who generally live in urban areas and/or who are generally working/lower-middle-class (Alkon and Agyeman 2011, Gottlieb and Joshi, 2010, Penniman 2018, White 2019). From their standpoints, food issues are not just about eating healthier foods, but about the structural inequalities based on race, social class, and gender that make it difficult to afford/access/grow healthier food (and at times, any food) (Guthman 2008, White 2019). For example, Karen Washington, a FJM activist, urban/rural farmer, and co-founder of the Black Urban Growers,[5] has spearheaded changing food terminology in relation to food deserts, stating that it "sugar coats" (Brones 2018, para. 13) the problems, whereas if you use her term "food apartheid" then "…a real conversation begins" (para. 13), as in looking at the intentional racist social structures that have created the food inequalities in the first place. In this manner, the focus of FJM is not just on food or on a given individual's relationship to food but on the social, economic, and political structures that shape systemic inequalities, of which food is a vital part along with jobs, housing, education, and healthcare. As the geographer Julie Guthman has insightfully observed, those who desire to engage in food activism should "… shift away from the particular qualities of food towards the injustices that underlie disparities in food access" and as a result, "… pay more attention to projects considered much more difficult in the current political climate; eliminating redlining, investing in urban renewal, expanding entitlement programs, obtaining living wages …" (2008, p. 443). I see co-creating and continuing to engage in developing CLiCK as this kind of 'difficult' structural work, in that CLiCK provides individuals of limited income in our community opportunities to create small locally based food businesses in a shared-use space. As such, CLiCK's members gain vital social and economic support, while they in turn are able to contribute to their own livelihoods and to our local food economy. Achieving this is more than merely addressing the 'qualities of food' that certain individuals can purchase: it seeks to create collective solutions aimed at changing people's relationships to different forms of knowledge, as well as material resources.

Attempting to embody a FJM perspective through CLiCK's mission has obviously been no easy task, both as a result of our positionalities and the existing social structures. Our social and economic systems are not set up to create co-operative, inclusive, shared-use, non-profit, locally based alternatives to global capitalism's business as usual (McLaren and Agyeman 2015). Through a model of collective equitable economic empowerment instead of an individually focused charity model of aid-based dependency, CLiCK has locally challenged social and economic inequalities in small ways that have not been universally welcomed by some members of our community, some of whom I consider part of the AFM. In fact, this has been an aspect of our struggle – one that I will critically analyze using intersectionality – that has shaped CLiCK's story on all social levels, micro as well as macro. However, again I do not, nor would I, ever claim that CLiCK has achieved a permanent practice of 'food justice' or even 'JS' (such a permanent state, I would argue, is not even possible). Rather, I uphold as the educator, socialist, and co-founder of the Highlander Folk School Myles Horton said, we "make the road by walking" (Bell et al. 1990), in that we can only contribute to creating 'justice' by collectively, consciously, and ultimately *reflexively* (DuPuis et al. 2011) working to create it through moment by moment *praxis*. If I were an outsider, an academic engaging in ethnographic research of CLiCK by relying on what is stated on our website and promotion materials, or in interviews and observations, I might have missed some aspects of its operations, especially the elusive complex micro engagements between individuals, as well as between other organizations and community groups that are often unperceivable to an outsider. Obviously, I have not witnessed all that has gone on at CLiCK, nor can I fully evaluate anyone's experiences other than my own, but I ask that readers trust my discernment and my evaluative abilities in terms of proposing answers to this book's questions. However, in the end, I gauge that it is not my place alone to make a final assessment of the possible answers that emerge from this narrative, even if such definitive conclusions were achievable. Rather, it is the role of others, including readers and perhaps ultimately it is up to posterity. Regardless, it is my hope that such answers, whatever form they take and for whomever they resonate, will have intellectual and social merit as part of the continual global project of imagining, creating, and recreating JS.

In returning to the declaration that this is 'my story', one now qualified as being told from the perspective of my positionality, I want to recognize the structural privileges embedded in being able to make a claim to one's story and in being able to tell it. These privileges based on race, social class, education, etc. grant me social status so that my story is recognized as being of social importance and interest. In my role as an academic, I have a full-time (although not tenure-track) position that gives me some time and the resources (a room of my own) to engage in intellectual activism and now to write. Of course, spending time over the last 11+ years in my 'intellectual activist' role at CLiCK has brought me minimal tangible academic rewards, but I still see myself as fortunate to be able to straddle the two worlds. In contrast, there are many others who are doing similar and/or more significant food justice/JS work, particularly in low-income communities and/or communities of color, who do not have such privileges and who are therefore not able or asked to tell their stories. I want to be cognizant of this fact, as even though their stories (past and present) may remain untold, they nevertheless remain connected to my story, in that CLiCK is just one small piece in a historically rooted and socially collective collage that seeks to turn justice from a noun into a living and vibrant verb, as declared by Judge Wendell Griffen in his blog titled, "Justice is a verb!" (Griffen 2018). This call for justice to be a living and vibrant verb – a verb of *praxis* – increases in relevance day by day in light of our ongoing environmental and social collapse (Baker 2013; Diamond 2005; Figueres and Rivett-Carnac 2020; Parenti 2012) characterized by increasing racial and social class inequalities created by the ongoing privatization of the 'commons' (Wade 2005). As such, my 11+-year ongoing CLiCK project has been, and remains, motivated by Plato's assertion that, "Knowledge without justice ought to be called cunning rather than wisdom".[6] In my experience, the academy mostly encourages and rewards the former, not the latter, and does so at the expense of many academics' 'original passion'. As someone who is non-tenure track, I see the many advantages to being a tenure-track/tenured professor but I also see the dangers of losing touch with one's heart. One of my colleagues astutely and surprisingly said to me once, "You chose the path of the heart, didn't you?" I unquestioningly answered, "Yes", although it would have been more honest to have said that it chose me, perhaps because 'it was a good idea'.

As a professional chocolatier who lived and worked in the Mid-Atlantic area for the first couple of years with my own private space I had to start from scratch when I moved to Connecticut.... I needed to find a commercial kitchen, a home, for my fledgling chocolate shop. I stumbled upon CLiCK...
CLiCK encouraged me to speak up about my needs as a 'foodpreneur'.

African American female kitchen member
I have stayed involved on the board because I am persistent and committed to the mission of CLiCK. I regard it as a haven of clear thinking and useful action in a larger context that can be challenging.
Middle-class white male board member

Storytelling methods

In cultural anthropology, the practice of engaging in 'insider' and 'outsider' ethnographic work originated from linguistic studies in the 1960s and is referred to as emic (from the insider's perspective, as told/shown to the anthropologist) and etic (from the anthropologist's outsider perspective about the insiders, based on observations and interpretations). In other words, the emic is the one observed, whose story is told, and the etic is the observer, the one who records, analyzes, and writes about the other. In my case, I am a hybrid of the emic and the etic, crossing between the two, which I see as embodying what Anna Louise Keating, in her book *Transformation Now!: Towards a Post-Oppositional Politics of Change* (2012), refers to as 'threshold theories', to underscore their nonbinary, liminal, potentially transformative status (p. 10). Keating uses the term, "*thresholds* [italics in original] to represent complex interconnections among a variety of sometimes contradictory worlds – points crossed by multiple intersecting possibilities, opportunities and challenges" (p. 10). Likewise, I am simultaneously engaging in an autoethnography, as well as a collaborative ethnography and a standard ethnography. Given my positionality within CLiCK, these methods are entangled, just as my roles as insider and outsider are ultimately inseparable from the roles others have played at CLiCK (Barad 2007) and CLiCK itself is ultimately inseparable from the patterns of the surrounding society, as will be further explored. Likewise, since this is a work of memory, of personal storytelling, and of collective oral history, the levels of entanglement will vary; the roots, however, will remain firmly within my perspective and my positionality. At the same time, branches will emerge from the roots through the use of many other materials, woven into my narrative, along with your *positionality* as the reader, your knowledge of the topics under discussion, your interpretations and objectives. As with CLiCK, the writing of this book and all the ways it will be read, and hopefully used, are collective, co-constructive endeavors.

Given that I still live with my wife and still work with all of the people who are also part of my story, I am not specifically interested in them as unique individuals, though the story demands I do offer some references to other's actions, but they are still seen as part of the social patterns mentioned above. Sociologist Max Weber coined the term 'ideal types', also known as 'pure types', representing abstract and hypothetical concepts that are useful in social analysis. An ideal type is not understood as being the perfect manifestation of the thing in question but is made up of a given thing's characteristics and elements, although not necessarily all of them. Weber mostly applied his ideal type analysis to macro social concepts such as 'the Protestant ethic' or 'capitalism', but it can also work for constructing 'ideal type individuals' as I will loosely do (Swedberg 2018). In taking some but not all of people's characteristics and elements, and in filtering them through my memories, as well as through an intersectional lens, I will be creating 'characters', who are based in reality on actual people but who through critical analysis become for the most part fictional. Weber recognized the 'fictional' aspect to ideal types and never claimed for them to correspond directly with social reality. Likewise, persons in this book are based on my real experiences and their real actions but have to a degree nevertheless been turned into ideal types, put into the kaleido-scope of my mind (my experiences intersecting with my sociological training), and placed into a larger analytical social context. I do this not only for the sake of social etiquette but also because what matters here are not the specific individuals involved (including me) but rather what our social identities and positionalities mean in relation to our actions and what together they can illuminate about interpersonal/group dynamics, about society, and ultimately about the answers to my questions and the goal of creating a more just and sustainable world. It is for this reason that quotes from those involved are scattered throughout the book (see the Appendix for the actual questions) and are left to speak for themselves as if they were in the room offering their input. In addition, although I am interested in my specific positionality, these quotes are attributed to 'ideal types' rather than linking their statements to specific people in relation to CLiCK. Finally, those who have been/are still involved with CLiCK may well recognize themselves in the story and/or their quotes, but it is my hope that they also recognize that I am not interested in them as fully complex and living 'individuals' but in the social roles they have played/continue to play.

The organization of this book is based on layering from theory to praxis and back again, so that readers may share in the same theoretical tools I will be using to critically evaluate CLiCK in practice, while I also weave in other illustrative case studies to offer material for some comparative analyses. Part I is 'Under-standing in Theory'. In Chapter 1, 'Understanding just sustainabilities in theory', I introduce the concept of 'just sustainabilities' (JS) while reviewing germinal literature both past and current. I show how theories can shape praxis, while reaffirming that practice is always in the moment and therefore remains unstable, variable, and ultimately raw until critically analyzed. I also review other JS case studies based on current literature for illustrative purposes. In Chapter 2, 'Under-standing intersectionality in theory', I introduce the concept of 'intersectionality' and follow the model of Chapter 1. In Chapter 3, 'Overlapping lenses: insights into praxis', I theorize the overlapping of JS and intersectionality and argue that together they provide a means of engaging in reflexive analysis to help guide 'practice' to be 'just', including ways of being anti-racist/anti-classist/anti-sexist. Additionally, I offer case studies to illustrate how the theoretical merging of just sustainability and intersectionality can lead to innovative social practices in rela-tion to local and/or global food systems.

PART II is 'Understanding from Within and Without in Practice'. In Chapter 4, 'Origins of CLiCK from within and without', I present the historical origins of CLiCK from an insider/micro auto-ethnographic perspective using my own recollections, as well as collective/ethnographic perspectives based on stories from other CLiCK insiders. I will use these primary resources, vignettes, and other case examples to illustrate the challenges faced in starting a shared-use commercial kitchen dedicated to food justice. I also present the historical origins of CLiCK from an outsider/macro perspective by examining outsider stories, the community context that shaped CLiCK's origins, and secondary sources that focus on local/

global food systems. These combined micro and macro perspectives create a holistic understanding of the specific connectivity between CLiCK's interpersonal dynamics and the inter-agency and intercommunity dynamics that have both helped and hindered CLiCK's origins and development. Chapter 5, 'Development of CLiCK from within and without', and Chapter 6, 'Institutionalization of CLiCK from within and without', both follow the same model as Chapter 4 but with the focus on the *development* of CLiCK and the current attempts to *institutionalize* CLiCK, respectively. Chapter 6, like the previous two chapters, continues the story of CLiCK's growth and acknowledges the struggles that food justice agencies/non-profits face in attempting to grow into established institutions without losing their ideological commitments to and practice of social/food justice. PART III is 'Successes, Failures, and Thresholds – An Exploration of Praxis'. In Chapter 7, 'Putting just sustainabilities and intersectionality into praxis?', I synthesize the theory from Part I with the practice from Part II and continue to tell CLiCK's story while highlighting steps we took to put our theoretical ideals into *praxis*. I explore the question as to 'what is community' and develop the proposed concept of the 'JS imaginary', giving examples of how I think we created it by making CLiCK *Real*, first to ourselves and second to the larger community. Chapter 8, 'Thresholds of success…', is my evaluation of CLiCK's 'successes', based on the lenses of JS and intersectionality in theory and practice, hence praxis, including how such successes link to the future of local and global food systems. I further recognize the internal/micro practices illustrated in previous chapters, including issues of racism, sexism, classism, and other manifestations of the vagaries of human interaction that are reinforced by external/macro dynamics, including the challenges of attempting to create progressive social change in a highly unequal and capitalist society. In Chapter 9, 'Failures and unknowns (thus far…)', I present a number of CLiCK's 'failures', which I argue are also rooted in the systemic failures of our society. This is not to remove CLiCK's responsibility to still attempt to achieve challenging goals, but rather to recognize that separating micro organizational failures from the macro societal ones is not in fact 'theoretically', or practically, possible. The hope, however, is that by openly analyzing our failures, other organizations seeking to explore practices of JS from an intersectional perspective may benefit.

The Conclusion, '*Interconnections* now and beyond…', is my final reflection on the journey of co-founding CLiCK, of being the board president for its first six years, and of writing a book about it. I offer lessons learned and propose insights to further advance the cause of not only *theorizing* JS and intersectionality, but also *practicing*, albeit imperfectly, on the personal and political, hence the micro and macro levels in relation to the industrial food system and society as a whole. I return to my questions posed at the beginning of the book and attempt to finally answer them in relation to our ongoing *praxis*. Finally, in the Appendix, I include the questions I sent out and a selection of shared-use kitchens that inspired CLiCK, as well as a final postscript on the theme of becoming *Real*.

It is my hope that through this book's unique combination of autoethnography, participant observation, interviews, individual statements, and secondary research, readers will gain insights into CLiCK's micro and macro successes, failures, and unknowns in relation to its attempt to put the concept of JS into practice. Additionally, I hope that readers interested in linking theory with practice, especially in relation to JS/intersectionality, will appreciate this book's theoretical basis, upon which its practical analyses are developed. I hope that others involved with social justice/food/sustainability movements will benefit from this book's insights into 'best practices' for addressing social inequalities on the micro level, as well as the macro intersectional analysis in relation to JS that is often missing from 'food movement' books aimed at more general audiences. As an academic book, I have, as per my training, included many citations throughout (including from my own previous publications) with the goal of not only sharing with readers a wealth of resources that have shaped my thinking and understanding, but also to fully embody the spirit and the act of *collaboration*, recognizing that the solutions to our failing society, hence food system, *must* come from a multitude of perspectives, voices, and disciplines including academics, activists, artists, and poets, both living and deceased.

Finally, once again, I must state that this is my story and so I end with the last stanza of Niebuhr's perspicacious quote: "No virtuous act is quite as virtuous from the standpoint of our friend or foe as it is from our standpoint. Therefore, we must be saved by the final form of love which is forgiveness" (1952, p. 63). Thus, to my friends and foes and to readers unknown, I ask 'forgiveness'. I ask you to remember that this is my attempt at expressing in words "…a loving commitment to community" (White 2019, p. 27) and to all that we have created, and can still create, together. However, I alone am responsible for any 'unvirtuous' errors, and for the ways they intersect and are ultimately inseparable from that which is also deemed as being 'good'.

What attracted me to CLiCK is it is one of the only kitchen co-op solutions available in the state…CLiCK has allowed me to rekindle my passion and start my catering business. I currently still operate out of CLiCK and also attend board meetings.
  Middle-class white male entrepreneur/kitchen and board member

Fundamentally, like almost all incubator-type operations, the hardest nut to crack is the cost of real estate. Successful incubators typically have a sponsor that provides the space – an educational institution, the government, or some other angel. Expecting the people you most want to help (lower-income residents) to be able to provide the financial support to operate the program AND pay the rent puts the leaders in an ongoing untenable position. But I have bought into the vision, the multiple benefits of such a venture (economic development, improving healthy food access, creating sustainable small businesses, cultural diversity), and the enthusiasm of the founders and their successors.
  Middle-class white female volunteer and legal advisor

Notes

1 https://millmuseum.org/home-again/more/swift-waters-the-industrial-environment/swift-waters-or-cedar-swamp/
2 www.clickwillimantic.com
3 https://www.karenthefarmer.com/about
4 https://civileats.com/2011/12/19/tft-interview-malik-yakini-of-detroits-black-community-food-security-network/
5 https://www.blackurbangrowers.org/
6 https://www.brainyquote.com/quotes/plato_117922

References

Acevedo, S. M., et al. 2015. 'Positionality as knowledge: From pedagogy to praxis', *Integral Review* 11(1): 28–46.

Agyeman, J. 2003. *Just sustainabilities: Development in an Unequal World.* Cambridge, MA: MIT Press.

Agyeman, J. 2013. *Introducing just sustainabilities: Policy, Planning and Practices.* New York: Zed Books.

Alkon, A. H. and Agyeman, J. eds. 2011. *Cultivating food justice: Race, class, and sustainability.* Cambridge, MA: MIT Press.

Baker, C. 2013. *Collapsing consciously: Transformative truths for turbulent times.* Berkeley, CA: North Atlantic Books.

Barad, K. 2007. *Meeting the universe halfway: Quantum physics and the entanglement of matter and meaning.* Charlotte, NC: Duke University Press.

Bell, B., Gaventa, J. and Peters, J. 1990. *We make the road by walking: Conversations on education and social change.* Miles Horton and Paulo Freire. Philadelphia, PA: Temple University Press.

Berry, W. 1977. *The unsettling of America: Culture and agriculture.* Berkeley, CA: Counterpoint Publishing.

Berry, W. 2009. *Bringing it to the table: On farming and food.* Berkeley, CA: Counterpoint Publishing.

Bradley, K. and Herrera, H. 2016. 'Decolonizing food justice: Naming, resisting and re-searching colonizing forces in the movement', *Antipode* 48(1): 97–114.

Brones, A. 2018. Karen Washington: It's not a food desert, it's food apartheid [online]. Guernica. Available at: https://www.guernicamag.com/karen-washington-its-not-a-food-desert-its-food-apartheid/ [Accessed September 22, 2020]

brown, a. m. 2017. *Emergent strategy: Shaping change, changing worlds.* Chico, CA: AK Press.

Burawoy, M. 2005 'For public sociology'. 2004 ASA Presidential Address. *American Sociological Review* 70 February: 4–28.

Collins, P. H. 2009. 'Foreword: Merging intersections – Building knowledge and transforming institutions'. In Dill, B. T. and Zambrana, R. E., eds. *Emerging intersections: Race, class, and gender in theory, policy, and practice.* Piscataway, NJ: Rutgers University Press, pp. vii–xiv.

Collins, P. H. 2013. *On intellectual activism.* Philadelphia, PA: Temple University Press.

Commoner, B. 1971. *The closing circle: Nature, man and technology.* New York: Random House.

Crenshaw, K. 1989. 'Demarginalizing the intersection of race and sex: A black feminist critique of antidiscrimination doctrine, feminist theory and antiracist politics', *University of Chicago Legal Forum* 1, Article 8. [online]. Available at: https://chicagounbound. uchicago.edu/cgi/viewcontent.cgi?article=1052&context=uclf [Accessed September 22, 2020]

Crenshaw, K. 1991. 'Mapping the margins: Intersectionality, identity politics, and violence against women of color', *Stanford Law Review* 43(6): 1241–1299.

De la Barca, P. C. 2008. 'Life's a dream / La vida es sueño'. Applebaum, S., ed. and trans. Mineola, NY: Dover Publications. Available at: https://quotepark.com/works/ life-is-a-dream-8419/

Diamond, J. 2005. *Collapse: How societies choose to fail or succeed.* New York: Penguin Books.

DuPuis, M., Harrison, J. L. and Goodman, D. 2011. 'Just food?.' In Alkon, A. H. and Agyeman, J., eds. *Cultivating food justice: Race, class, and sustainability.* Cambridge, MA: The MIT Press, pp. 47–64.

Figueres, C. and Rivett-Carnac, T. 2020. *The future we choose: Surviving the climate crisis.* New York: Knopf.

Godfrey, P. 2017. 'Reflexive food-truck justice: A case study in CLiCK, Inc, a non-profit shared-use commercial kitchen'. In Agyeman, J., Matthews, C. and Sobel, H., eds. *Food trucks, cultural identity, and social justice: From loncheras to lobsta love.* Cambridge, MA: MIT Press, pp. 149–165.

Godfrey, P. and Freake, H. 2016. 'Feeding community: A case study of a shared-use commercial kitchen in eastern Connecticut'. In Bosso, C., ed. *Feeding cities: Improving local food access, sustainability, and resilience.* London: Routledge, pp. 113–128.

Godfrey, P. and Torres, D., eds. 2016a. *Systemic crises of global climate change: Intersections of race, class and gender.* London: Routledge.

Godfrey, P. and Torres, D. 2020. (Situational Strangers) Recipes for immigrant lives: Crossing, cooking, cultivating and culture at a shared-use commercial kitchen. In Agyeman, J. and Giacalone, S., eds. *The immigrant food nexus: Borders, labor and identity in North America.* Boston, MA: MIT Press, pp. 281–298.

Gottlieb, R. and Joshi, A. 2010. *Food justice.* Cambridge, MA: MIT Press.

Greenfield, P. 2020. 'World fails to meet a single target to stop destruction of nature – UN report'. [online] *The Guardian.* Available at: https://www.theguardian.com/environment/2020/ sep/15/every-global-target-to-stem-destruction-of-nature-by-2020-missed-un-report-aoe [Accessed September 22, 2020]

Griffen, W. 2014. *Justice is a verb!* [online]. Available at: http://wendelllgriffen.blogspot. com/ [Accessed September 22, 2020]

Guthman, J. 2007. '"If they only knew": Color blindness and universalism in California alternative food institutions'. *The Professional Geographer* 60(3): 387–397.

Guthman, J. 2008. 'Bringing good food to others: Investigating the subjects of alternative food practice'. *Cultural Geographies* 15(4): 431–447.

Keating, A. 2012. *Transformation Now!: Towards a Post-Oppositional Politics of Change.* Champaign, IL: University of Illinois Press.

McLaren, D. and Agyeman, J. 2015. *Sharing cities: A case for truly smart and sustainable cities.* Cambridge, MA: MIT Press.

Mills, C. W. 1959. *The sociological imagination.* Oxford: Oxford University Press.

Milner, H. R. 2007. 'Race, culture, and researcher positionality: Working through dangers seen, unseen, and unforeseen', *Educational Researcher* 36(7): 388–400.

Lappe, F. M. 1971. *Diet for a small planet.* New York: Ballantine Books.

Lassiter, L. E. 2004. 'Collaborative ethnography', *AnthroNotes* 25(1): 1–20.

Nestle, N. 2006. *What to eat.* New York: North Point Press.

Niebuhr, R. 1952. *The irony of American history*. New York: Charles Scribner's Sons.

Parenti, C. 2012. *Tropic of chaos: Climate change and the new geography of violence*. New York: Nation Books.

Patel, R. 2012. *Stuffed and starved: The hidden battle for the world's food system*. New York: Melville House.

Pellow, D. 2016. 'Toward a critical environmental justice studies: Black Lives Matter as an environmental justice challenge'. *Du Bois Review* 13(2): 1–16.

Penningman, L. 2018. *Farming while black: Soul Fire Farm's practical guide to liberation on the land*. Boston, MA: Chelsea Green Publishing.

Pollan, M. 2006. *The Omnivore's dilemma: A natural history of four meals*. London: Penguin Books.

Rose, G. 1997. 'Situating knowledges: Positionalities, reflexivities, and other tactics', *Progress in Human Geography* 21(3): 305–320.

Schlosser, E. 2001. *Fast food nation: The dark side of the all-American meal*. Boston, MA: Mariner Books.

Shiva, V. 2000. *Stolen harvest: The hijacking of the global food supply*. Boston, MA: South End Press.

Slocum, R. 2007. 'Whiteness, space and alternative food practice', *Geoforum* 38(3): 520–533.

Smith, D. 1990. *The conceptual practices of power: A feminist sociology of knowledge*. Toronto: University of Toronto Press.

Stebbins, S. 2019. 'Poorest town in every state'. [online] *24/7 Wall St.* Available at: https://247wallst.com/special-report/2019/05/01/poorest-town-in-every-state-5/3/ [Accessed March 18, 2020]

Swedberg, R. 2018. 'How to use Max Weber's ideal type in sociological analysis', *Journal of Classical Sociology* 18(3): 181–196.

UN News, 2020. 'Rising inequality affecting more than two-thirds of the globe, but it's not inevitable: new UN report.'[online]. Available at: https://news.un.org/en/story/2020/01/1055681 [Accessed March 18, 2020]

Wade, B. 2005. 'A new tragedy for the commons: The threat of privatization to national parks (and other public lands)', *The George Wright Forum* 22(2): 61–67. [online]. Available at: http://www.georgewright.org/222wade.pdf [Accessed March 18, 2020]

White, M. 2019. *Freedom farmers: Agriculture resistance and the black freedom movement*. Chapel Hill: The University of North Carolina Press.

Part I
Understanding in theory

1 Understanding just sustainabilities in theory

Justice cannot be for one side alone, but must be for both.

Eleanor Roosevelt

Musings on the concept of justice and my personal calling

A life challenge for those who seek a more 'just' world, like myself, involves agreeing upon a theoretical answer to the question, 'what is justice?', in terms of its "distribution, inclusion, participation, recognition, fairness / procedure and capabilities" (Pellow 2018, p. 11). And, if such an answer were to be agreed upon, even more challenging would be finding answers to the related questions: Now that we have defined justice, how is it to be achieved, and if achieved, what then does it look like, on what levels of scale can it manifest, and how is it to be sustained? And then what might it feel like and for whom in terms of individuals, homogeneous groups, such as organizations, and/or highly pluralistic groups such as nations? These are some of the questions (there are many more...) whose answers I have been seeking (as have many others...) over the last 30 years, during which time I have identified myself as a social justice activist and later as a social justice academic. Of course, how a person answers depends significantly on their positionality (Alkon and Agyeman 2011; Rose 1997) and the extent to which they (individually and ultimately collectively) and their loved ones are able to meet their cultural, physical, emotional, and spiritual 'needs'. Of course, such a nebulous word as 'needs' leads to further discussion and debate as to what and how much people need, often confused in our own culture with what we *want*. Linking the idea of needs with what has been referred to as the "human dignity line" (Larrain et al. 2003, in Agyeman 2012) adds clarity, as shelter, food, water, and culture can be understood as being necessary for human dignity. This culturally specific 'line' is similar to our own poverty line here in the USA, in that it marks a level at which or above which one is considered able to achieve sufficient material consumption; I would add, as does the Universal Declaration of Human Rights (UDHR 1948), that cultural expression and numerous other freedoms beyond consumption are also essential for a life of dignity. However, as indicated, 'needs' are relational and culturally specific, making their assessment complex as well as intersectional (as will be discussed in Chapter 2), and they are intimately linked to conceptions of justice. As a result, conceptions of justice, whether empirical (based on how people evaluate real events) or normative (based on what people think ought to exist derived from moral and/or ethical arguments), are nevertheless highly contested (Kluegel et al. 1995). In fact, their threads travel across and through the dynamically unequal social landscapes of nation-states and cultures, as well as through collective and individual identities shaped by social class, race, ethnicity, gender, and other socially determinant criteria embedded in existing power structures. Hence, there is little agreement among justice theorists as to how justice should be defined (Brighouse 2004; Capek 1993; Pellow 2018; Schlosberg 2007; Sen 1990, 2008, 2012), and even if there were agreement, we would still be left questioning how best to achieve and sustain it so we may individually and more importantly *collectively* feel it in practice (Sen 2008). Nevertheless, when it comes to justice in relation to food, the concept of right and wrong is irrelevant as Pearl S. Buck once said that, "A hungry man can't see right or wrong. He just sees food".[1] As such, at its most fundamental level, justice must include allowing all people, or better yet all beings, access to sufficient food, water, and shelter with which to at a minimum live with dignity and perhaps to even thrive in their lives. Therefore, the primary justice issue at work in a person stealing food is not in the theft itself (or as is most likely the case, eating that which has been declared waste by a wasteful system), but rather in the initial inequality that created the disempowered hunger, hence the need for the theft in the first place.

The spiritual philosopher Osho shares a story about the Chinese philosopher Lao Tzu who was asked by the then-Emperor to be his chief of the supreme court. Lao Tzu argued he was the wrong man for the job as he believed that "…the system is wrong" (Osho 2000, p. 4), but still the Emperor insisted. When later asked to judge a thief, who had stolen from the richest man in the city, Lao Tzu concluded that both should go to jail, even though he believed this outcome to be unfair to the thief. To the indignant rich man, Lao Tzu explained, "Your need to be in jail is greater, because…Your very greed is creating these thieves. You are responsible. The first crime is yours" (p. 4). The Emperor unsurprisingly soon relieved Lao Tzu of his position; as he left, he said, "I told you I am not the right man. The reality is that your society, your law, and your constitution are not right. You need wrong people to run this whole wrong system" (p. 5). Taking Lao Tzu's perspective, it can again be argued that a social system that denies people necessities based on their economic standing or lack thereof can never be seen as right or *just*. I have joked with my students that Jesus' meal of loaves and fishes would have a very different resonance had he asked for payment and turned away those unable to pay, even though other authority figures charging for meals at the time would not have seemed amiss. Hence, the concept of justice and any attempts at its practice have been and remain highly contested and controversial, with ideas of 'right' and 'wrong' heavily shaped by social positions and broader societal systems.

Ironically, the emphasis on equal access to resources isn't just something spiritual teachers seek to preach and practice for their own moral and ethical reasons. It turns out that more equal societies greatly improve many life experiences for *all* members of a given society. In the book *The Spirit Level: Why More Equal Societies Almost Always Do Better* (2009), Pickett and Wilkinson argue that the "...quality of social relations in a society is built on material foundations. The scale of income differences has a powerful effect on how we relate to each other" (p. 4). As a result, they highlight correlations between social and economic equality within a given society and a whole slew of other social and environmental factors such as obesity, mental illness, teen pregnancy, and even recycling, noting that more equal societies have lower rates of the unfavorable outcomes and higher rates of the more favorable outcomes. Of particular interest here are Pickett and Wilkinson's findings on recycling, which are supported by other studies that make links between social equity and pro-environmental laws and behaviors (Boyce 1994; ; Rao and Min 2018; Torras and Boyce 1998). This research affirms Julian Agyeman's assertion that "human inequality is bad for environmental quality" (2008, p. 752), as well as, unsurprisingly, bad for *all* other social outcomes.

Given that equality is seemingly so essential to human well-being, another mystery that I have pondered is why some people are drawn to what Martin Luther King Jr. referred to as the "...hard sacrificial, suffering inducing struggles..." (King 2010) involved in social justice work, while others (unfortunately) are not. Of course, for most people, there is an affinity for some form of justice, at least as understood from their own social position, in perhaps a more 'normative' than 'empirical' manner. However, as Eleanor Roosevelt recognizes, such one-sidedness merely from our own purview begs the question if there is justice at all, even though claims may be made to the contrary (as the elite tend to do).[2] For others, the social injustices that shape their everyday existences are so overwhelming that they may feel unable to struggle for social change, as the effort poses too great a challenge to their basic physical survival, while still others in similar situations may find that *not* struggling inversely poses too great a challenge to their basic spiritual and physical survival. In such cases, individuals and/or groups may seek to redress such injustices, efforts that will no doubt be confrontational to varying degrees, as they will involve proposing and ultimately attempting some form of redistribution and/or retribution. Such acts may be initiated by members of the dominant group who nevertheless seek to act in just ways (as illustrated by Lao Tzu), or they may also be taken on by the ones experiencing the injustice in the form of social movements (as illustrated by Jesus). Overall, social movements around the world are dependent on three key factors – political opportunity, mobilizing structures, and framing (McAdam et al. 2012) – with the variability of each factor influencing a movement's success and/or failure. Demands made by a given social movement generally embody some collective definition of justice (at least from the movement's perspective) and can involve policy changes (for example, the feminist movement and civil rights movement), forms of redistributive justice (for example, ongoing struggles for reparations, the Equal Rights Amendment [ERA], and for increasing the minimum wage), and/or regime change (for example, the Cuban Revolution and other successful revolutions, as well as more recently, the 2019 protests in Puerto Rico that resulted in the governor stepping down). On the other hand, such movements, to the degree that they are successful and hence seen as political, social, and economic threats to the status quo, may be met with increased *injustices* in the form of political disenfranchisement (for example, Jim Crow on the tail of Reconstruction after the Civil War), resource deprivation (for example, imprisonment and other forms of punishment), and/ or military suppression (for example, the ongoing struggles by the Palestinians for justice against Israeli occupation and military oppression). Of course, not all social movements are politically progressive but all seemingly cling to some vague concept of justice, as in, for example, pro-lifers seeking justice for the unborn (Hekman 1984). However, in my opinion, only progressive social movements actually embody the practice of justice for *all*, even as the status quo uses and abuses it by appealing to its universally seductive symbolism. Such progressive struggles always evoke (or *should*, if they be genuine) what Adelina Mkami (2020) refers to as the "thorns of justice" (p. 113), capable of discomfiting those with privilege, as well as challenging advocates by holding them accountable to embody and practice their own claims.

I have always been drawn to these 'thorns'. In fact, I have never doubted that my emotional and spiritual *need* to be involved in the struggle for justice has been central to my personal 'calling'. I am specifically using the term 'calling' in honor of Max Weber, who in his *Protestant Ethic and the Spirit of Capitalism* (1905) put forth the notion that the Puritan was attuned to a desired calling as a means to serve 'the Lord'. In contrast, we under capitalism are now forced to a seemingly inauthentic calling, and as such find ourselves trapped in what Weber infamously referred to as the 'iron cage', the seductive reduction of all our social relations down to their most base expressions through monetary exchanges. Yet, despite having been forced, like most everyone else, to find a vocation in exchange for wages, I nevertheless have been able/chosen (the line between personal/social ability and choice is and should remain unclear) to hold onto the more historical, seminal, and spiritual notions of a 'calling'. It is for this reason that I was not only drawn to what Martin Luther King Jr. referred to as the "...*tireless exertions and passionate concern*..." (King, 2010) involved, in this case, in co-founding and leading CLiCK (Commercially Licensed Co-operative Kitchen), but also why, when it came to theorizing the larger social implications of what CLiCK has come to mean, I was drawn to the works of Agyeman et al. (2003, Agyeman 2008) and his concept of JS.

> *I think the low-cost serv-safe classes and community education/events that are encouraging and welcoming to Spanish-language speakers definitely help promote social justice in the Windham area.*
>
> Middle-class white female volunteer/graduate student who worked with CLiCK for a yearlong multicultural food story writing workshop and final performance

Environmental justice, the emergence of just sustainability, and case studies

Originally devised by Julian Agyeman et al. (2003) and continuously developed independently by Agyeman and others, JS seek to reduce the disjuncture between the theory and practice of the concept of sustainability, as well as to ameliorate the failure to address issues of equity and justice in supposedly sustainable endeavors. The preeminent definition of sustainability by the World Commission on Environment and Development (WCED) was articulated in *Our Common Future*, also known as the Brundtland Report, and reads as follows: "...sustainable development is development that meets the needs of the present without compromising the ability of future generations to meet their own needs" (World Commission on Environment and Development 1987). Although this report marked the beginning of global attention to the concept of sustainability, its simplicity and vagueness have nevertheless left the proposed concept of sustainability itself open to ineffectiveness (Connelly 2007; Lélé 1991). As is the case with the concept of justice, the concept of sustainability suffers from a lack of clearly articulated and agreed upon definitions despite the WCED's noble attempt. As Agyeman recognizes, in the WCED's definition, there is little clarity over "what is to be sustained, by whom, for whom and what is the most desirable means of achieving this [these] goals" (Agyeman 2012, para. 3). Significantly absent from the WCED's definition of sustainability is any reference to 'justice' or 'equity' in relation to how resources are, or are not, shared to meet the aforementioned 'needs'. In contrast, Agyeman et al. (2003) define 'just sustainabilities' as "the need to ensure a better quality of life for all, now and into the future, in a just and equitable manner, whilst living within the limits of supporting ecosystems" (p. 2), thereby attempting to address what the WCED failed to recognize. Furthermore, Agyeman's later development of the plural use of the term 'sustainabilities' sought to recognize the diversities of possibilities based on culture and place-based practices (Agyeman 2013), as opposed to the homogenizing tendencies of the previous singular term. This plurality invites cultural relativism, as what is seen as just or sustainable in one place may not be seen as such in another or as by one versus another. Such flexibility is essential to ensure the outcome of both 'justice' *and* 'sustainability', while the broader framework of JS and its four corresponding principles help ensure a continuity of meaning. The four principles are (1) guiding change to improve our quality of life and well-being; (2) increasing intergenerational equity (meeting the needs of both present and future generations); (3) furthering justice and equity in process, procedure, and outcome; and (4) increasingly living within ecosystem limits (Agyeman 2013; Agyeman et al. 2003), all of which will be *indirectly* evoked though the evolving narrative, as opposed to being explicitly analyzed. In addition, as *principles*, not rigid criteria, they will be engaged with in relation to Keating's (2012) notion of 'threshold', as in being much more fluid and complex and therefore harder to evaluate in any definitive way.

Weaving back to my question about the nature of justice, it is important to note that Agyeman et al.'s linkage of sustainability with the concept of justice can trace its roots to the Environmental Justice (EJ) movement, also known as the EJ Paradigm (Taylor 2000). This movement emerged from the grassroots work of primarily disenfranchised communities of color addressing the intersections of racism and environmental degradation, as articulated in the landmark 1987 United Church of Christ study, "Toxic wastes and Race in the United States" (Commission for Racial Justice 1987). This seminal study articulated the clear and recognizably *intentional* dumping of toxic wastes in communities of color and sought to begin organizing in outrage against what became known as environmental racism. As Robert Bullard, known as the father of EJ, astutely observed,

> People of color have *always resisted* actions by government and private industry that threaten the quality of life in their communities. Until recently, this resistance was largely ignored by policymakers. This activism took place before the first Earth Day in 1970; however, many of these struggles went unnoticed or were defined as merely part of the "modern" environmental movement.
>
> (1994, p. 3) [italics added]

Additionally, the work of Dorceta Taylor, along with others working from ecofeminist perspectives (Warren 1997), made key links between environmental injustices and issues of gender, race, and class (Taylor 1997a, 1997b). Then, in 1991, 'The People of Color Environmental Leadership Summit' led to the adoption of 17 'Principles of Environmental Justice' that "have strengthened cross-racial and cross-sector alliances" (The Second National People of Color Environmental Leadership Summit 2002) while also helping, at least in theory, to define what is understood and meant by the term environmental justice. The Preamble begins,

> We, People of Color, gathered together at this multinational People of Color Environmental Leadership Summit, to begin to build a national and international movement of all peoples of color to fight the destruction and taking of our lands and communities, do hereby re-establish our spiritual interdependence to the sacredness of our Mother Earth.
>
> (Principles of Environmental Justice, 1991)[3]

The declaration then goes on to list the 17 principles, the last of which ends by pledging to act in such ways as to, "… ensure the health of the natural world for present and future generations" (Principles of Environmental Justice, 1991). As such, with its emphasis on these overarching principles, EJ's strength was (and still is) that it addressed what had "…largely been left out of the traditional environmentalism and environmental policy", while also newly conceptualizing "the word environment", as well as linking it to "the potent issues of justice, equity, and rights" (2002, p. 13). Previously, environmentalism was generally understood, and still is by the white status quo, to be about wilderness and wildlife, reflected in the work of the Audubon Society or Sierra Club – not about such issues as, for example, air quality in urban areas (Mitchell and Dorling 2003), uranium dumping on Native American reservations (Brook 1998), and "Dumping in Dixie" (Bullard 2008). This narrow understanding leads to some striking and most often very racist omissions. Indeed, as the controversial authors and co-founders of the Breakthrough Institute,[4] Shellenberger and Nordhaus (2005), have astutely and cuttingly asked about the US environmental movement, "Why, for instance, is a human-made phenomenon like global warming-which may kill hundreds of millions of human beings over the next century-considered 'environmental'? Why are poverty and war not considered environmental problems while global warming is?" (2005, p. 12; quoted in Agyeman 2008, p. 751). Of course, given their prioritization of 'technological solutions' over social ones, their means of addressing such questions are very different from JS, but nevertheless the increasing recognition of this inseparability between "environmental quality and human equality" (idem) can be linked to the conceptual shifts resulting from EJ, a powerful lens through which to critically analyze environmental *injustices*. These analyses, of course, must also recognize positionality, as in the intersections of race, class, gender, and other social inequalities (as will be explored more in Chapter 2) that take place in a multiplicity of environments both nationally and internationally (Cutter 1995).

From the perspective of EJ, the WCED definition's lack of emphasis on issues of justice and equality is alarming, although not surprising. Additionally, although only the last EJ principle explicitly mentions 'the future', the overall emphasis of the collection of principles on asserting 'our spiritual interdependence to the sacredness of our Mother Earth' constitutes for me the essence of sustainability – put plainly, there can be no sustainability of any significance unless we globally act upon this 'spiritual interdependence' in just and equitable ways.

In questioning how to do this, as in moving "...towards more sustainable futures", Agyeman and Warner have recognized the need to broaden the scope of EJ and, "... to find other ways to ground, structure and guide our analysis" (2002, p. 20), as well as, I would add, 'our practices'. In addition to centering issues of justice and equality in relation to our present and future relationships with each other, other beings, and Mother Earth, JS also seek to address the role of human agency in "place-building", which "...emphasizes the very human dynamics involved in the physical and social construction of places" (Agyeman and Warner 2002, p. 20). In bringing the lens of JS to place-building, including the more common notion of 'development', the goal is to position "equity issues at the forefront" (p. 23), as opposed to overlooking their saliency on the immediate local and regional levels, as well as the metropolitan and even national and international levels. Yet, at the time Agyeman and Warner were writing in 2002, it seems such goals had not been significantly achieved, save for a few small-scale examples; Agyeman and Warner's purpose was to offer "...a proposed research framework..." (p. 25), still in theoretical stages, wherein environmental justice and sustainability principles might intersect. Since then, Agyeman's work has continued to explore potential examples, most recently in relation to food with a number of edited volumes (2013) to which I have twice contributed (Godfrey 2017, 2020).

And yet, despite this emerging literature exploring actual examples of JS in practice both domestically and internationally, including this work, I will also be introducing to my analysis and to the example of CLiCK what Julia Cidell refers to as the "sustainable imaginary". Cidell defines the 'sustainable imaginary' as "...a society's understanding and vision of how resources are being used and should be used to ensure socio-environmental reproduction" (p. 169). Furthermore, as with the seminal work by Benedict Anderson, *Imagined Communities: Reflection on the Origins and Spread of Nationalism* (1983), she states that such, "Imaginaries are not imaginary'" (p. 169) but rather are social schemas, ideologies through which our understandings of the world are socially constructed, as explored. For my purposes, I will be amending Cidell's concept to make it the 'JS imaginary', thereby recognizing both the emphasis on 'justice' and on the plurality of *intersectional* narratives in relation to understanding JS. Additionally, my intention in doing so is to recognize that, as stated, both justice and sustainability, hence JS, are not and cannot be completed destinations but rather processes in *praxis* that continually require collective reflection, reassessment, and imagining.

Aspects of Agyeman and Warner's (2002) proposed framework that are helpful for my purposes here, and my larger project with CLiCK, include their recognition of "Building community capacity—developing the capacity of stakeholders and public functionaries to engage and collaborate in planning local practices for sustainability and environmental justice" (p. 25). To give an idea of what such practices could look like, they offer the example of "Neighbors Building Neighborhoods" (NBN) from Rochester, New York, which they identify as "...a model for participatory planning that makes racial and economic diversity a top priority" (p. 25). Their positive assessment of NBN was later supported by Min Fen Kooi, who in 2006 reported that NBN's Asset-Based Community Development (ABCD) process used novel methods that "increased citizen participation in neighborhood activities", resulting "... in numerous benefits to the city" (p. 66). As a result, according to Kooi, "...the program [became] nationally regarded as a model of 'Best Practices', which has inspired other cities, including Syracuse, NY, Des Moines, IA, and Newark, NJ" (pp. 66–67). However, Kooi notes that despite this apparent success, NBN continued to face several challenges, including the city government's ultimate control in decision making and project implementation that undermined citizen input (p. 60). At the time of Kooi's writing in 2006, NBN's future was uncertain due to mayoral changes in Rochester; based on my current research, it seems to have not been continued, even though, as Kooi recognized, it had not yet fully achieved its goals. I have chosen to mention this because often when reading exemplary case studies, it turns out that, as in this case, despite being illustrative of a researcher's argument, the case subjects may no longer exist a few years later. This lack of *sustainability* may actually challenge an author's original analysis and/or invite them or others to engage in further inquiry as to what happened and why such a project, venture, organization, etc. was discontinued. Furthermore, when a case study is illustrative of *progressive* social change, as again in this case, it is worth examining the relationship between the discontinuation of the particular project and the very attributes that made it socially and politically significant in the first place. For one of the ongoing central challenges we have faced at CLiCK has been its very survival and one of my struggles in writing this book has been my choice *not* to focus my time and energy on telling CLiCK's story if it meant, as it did up until recently, putting CLiCK's very survival at risk. In fact, since beginning this book, there have been numerous direct challenges to CLiCK's survival (such as struggles to pay the mortgage, as well as other monthly bills) which have forced me to stop writing and get back to doing the actual practice work of helping to keep CLiCK afloat. A number of these examples will be discussed in later chapters, but suffice it to repeat here what Lao Tzu said, '…the system is wrong', which makes doing that which is 'right' very challenging. Additionally, since our goal with CLiCK was (is) to create a *place* where social justice could be practiced, I have come to recognize, as the geographer Gilmore (2002) has argued, that "A geographic imperative lies at the heart of every struggle for social justice; justice is embodied, it is then therefore always spatial, which is to say, part of a process of making a place" (p. 16). And a large part of this struggle for so many, including CLiCK, involves the financial upkeep of a place, including the maintenance of land, property, and other necessary services, as will be discussed.

For another early example of JS in practice, Agyeman and Evans (2004) traced the term's discursive emergence in Britain, expressing some hope that dialogue around the concept of JS was then emerging among "progressive NGO's, academics and local community organizations worldwide" (p. 163). As a result, they proposed in their conclusion that the next step should be

> …for governments at the local, regional, national and international levels to learn from these environmental justice and progressive, or JS-based, organizations and to seek to embed the central principles and practical approaches of just sustainability into sustainable development policy.
>
> (p. 163)

To determine whether or not this identified need has been addressed since Agyeman and Evans proposed it back in 2004, Broto and Westman (2017) conducted extensive database research. Posing the research question, "To what extent are just sustainability principles integrated in a sample of flagship sustainability initiatives in cities and urban regions?", Broto and Westman reviewed 400 sustainability initiatives (both public – civil and governmental – and private) from 225 world cities. Under their conditions, all four principles had to be "…met simultaneously for an initiative to advance 'just sustainability'" (pp. 637–638). They used three levels of criteria ("addressed directly", "addressed indirectly", and "not addressed") for each of the four principles and ultimately found that overall "…just sustainability principles are *not* yet widely integrated in mainstream discourses of urban environmental planning" (p. 648). Many organizations did have aspects of one or two of the principles but, as Broto and Westman noted, this does "...not necessarily challenge the fundamental structures of social organization and knowledge production that produce injustices in the first place" (p. 648). Thus, they further concluded, echoing ideas already discussed, that to genuinely achieve JS *in practice* requires moving "...away from comfort zones and received environmental policy wisdom..." to intentionally engage in the "...difficult and rare..." work of "...redistributive efforts alongside strategies to address recognition struggles" (p. 648). Their final paragraph is eloquent and valuable, and I quote it here in full:

> Just sustainabilities is not a ready-made recipe to deliver concrete initiatives, but a set of principles that should guide, rather than dictate, action. Just sustainabilities is a discourse of hope. Its objective is to deliver discursive tools that can be appropriated by different actors to inspire visions of future sustainable and just cities and make them, or at least part of them, happen.
>
> (p. 648)

Obviously, I entirely agree. My work with CLiCK and this book has been motivated both by JS' 'discourse of hope' and by the 'discursive tools' it has given me and others to not only visualize possibilities for a more just and sustainable food system, but to more challengingly 'make them, or at least part of them, happen'. For theories do and should shape practice (and vice versa), hence *praxis*, and it is for this reason that I assert, as others have also done, that justice work, and in this case JS work, requires not only 'sacrifice, suffering, and struggle', but also needs the additional theoretical discourse of intersectionality and its subsequent practices, as will be explored in the next two chapters. Echoing back to Niebuhr and his insight that "Nothing that is worth doing can be achieved in our lifetime", I also propose here that justice and its achievement, as well as the *sustaining* of JS, requires, as stated, ongoing reflection, reassessment, and imagining over what the French Annales School of historical writing called the "longue durée" (the long term). For as the feminist physicist Karen Barad astutely states in her book *Meeting the Universe Halfway: Quantum Physics and the Entanglement of Matter and Meaning* (2007),

> Justice, which entails acknowledgment, recognition, and loving attention, is not a state that can be achieved once and for all. There are no solutions; there is only the ongoing practice of being open and alive to each meeting, each intra-action, so that we might use our ability to respond, our responsibility, to help awaken, to breathe life into ever new possibilities for living justly. The world and its possibilities for becoming are remade in each meeting.
>
> (p. ix)

And I would add, with each 'intersection', as will be explored in the next chapter.

> *I was drawn to serve on the CLiCK board because of the mission of driving local sustainable economic development through a shared kitchen. In particular, the idea of connecting small businesses with local farmers.*
> Middle-class white female former board member
>
> *As a recent immigrant myself, I was searching for ways to connect with the community where I live, motivated by the political tensions we were suffering from at the time and the stigmas that Hispanics were being labeled with by our political leaders…*
> *I knew that CLiCK was the place where I had to be to help the community that opened its arms and welcomed me so warmly. What attracted me the most was the transparency and leadership and the people who work at CLiCK. It is a place where people can find a way to start their own business being in charge of their own schedules and without risking their economic welfare regardless of their race, cultural background, ethnicity, native language, or gender. Moreover, people that reach out to CLiCK have a group of people who will guide them on the right path to be a successful and independent businesses.*
> Middle-class Latino board member

Notes

1 https://www.azquotes.com/quote/556964
2 https://www.goodreads.com/quotes/845081-justice-cannot-be-for-one-side-alone-but-must-be
3 https://www.nrdc.org/sites/default/files/ej-principles.pdf
4 https://thebreakthrough.org/

References

Agyeman, J. and Warner, K. 2002. 'Putting 'just sustainability' into place: From paradigm to practice', *Policy and Management Review* 2(1): 8–40.

Agyeman, J., Bullard, R. D. and Evans, B. 2003. 'Joined-up thinking: Bringing together sustainability, environmental justice, and equity'. In Agyeman, J., Bullard, R. D. and Evans, B., eds. *Just sustainabilities: Development in an unequal world.* Cambridge, MA: The MIT Press, pp. 1–16.

Agyeman, J. and Evans, B. 2004. "Just sustainability': The emerging discourse of environmental justice in Britain?', *The Geographical Journal* 170(2): 155–164.

Agyeman, J. 2008 'Toward a 'just' sustainability?' *Continuum – Journal of Media and Cultural Studies* 22(6): 751–757.

Agyeman, J. 2012. 'Just sustainabilities' [online]. Available at: https://julianagyeman.com/2012/09/21/just-sustainabilities/ [accessed September 23, 2020) [Accessed September 10, 2019].

Agyeman, J. 2013. *Introducing just sustainabilities: Policy, planning, and practice.* London and New York: Zed Books.

Alkon, A. H. and Agyeman, J., eds. 2011. *Cultivating food justice: Race, class, and sustainability.* Cambridge, MA: The MIT Press.

Anderson, B. 1983. *Imagined communities: Reflection on the origins and spread of nationalism.* London: Verso Books.

Barad, K. 2007. *Meeting the universe halfway: Quantum physics and the entanglement of matter and meaning.* Durham, NC and London: Duke University Press.

Boyce, J. K. 1994. 'Inequality as a cause of environmental degradation', *Ecological Economics* 11: 169–178.

Brighouse, H. 2004. *Justice.* Cambridge, MA: Polity Press.

Brook, D. 1998. 'Environmental genocide: Native Americans and toxic waste', *The American Journal of Economics and Sociology* 57(1): 105–113.

Broto, V. C. and Westman, L. 2017. 'Just sustainabilities and local action: Evidence from 400 flagship initiatives', *Local Environment* 22(5): 635–650.

Bullard, R. D. 1994. Unequal protection: Environmental justice and communities of color. San Francisco: Sierra Club Books.

Bulland, R. D. 2008. *Dumping in Dixie: Race, class and environmental quality.* New York: Avalon Publishing-(Westview Press).

Capek, S. 1993. 'The "environmental justice" frame: A conceptual discussion and an application', *Social Problems* 40(1): 5–24.

Commission for Racial Justice. 1987. 'Toxic wastes and race in the United States: A national report on the racial and socio-economic characteristics of communities with hazardous waste sites' [online]. Available at: https://www.nrc.gov/docs/ML1310/ML13109A339.pdf [Accessed September 10, 2019].

Connelly, S. 2007. 'Mapping sustainable development as a contested concept', *Local Environment* 12(3): 259–278.

Cutter, S. 1995. 'Race, class and environmental justice', *Progress in Geography* 19(1): 111–122.

Gilmore, R. W. 2002. 'Fatal couplings of power and difference: Notes on racism and geography,' *The Professional Geographer* 54(1): 15–24.

Godfrey, P. 2017. 'Reflexive food-truck justice: A case study in CLiCK, Inc, a non-profit shared-use commercial kitchen'. In Agyeman, J., Matthews, C. and Sobel, H., eds. *Food trucks, cultural identity, and social justice: From loncheras to lobsta love.* Cambridge, MA: MIT Press, pp. 149–165.

Godfrey, P & Torres, D. (May, 2020) 'Recipes for immigrant lives: Crossing, cultivating, cooking, and culture at a shared-use commercial kitchen'. In Giacalone, S. & Agyeman, J., eds. *The Immigrant-Food Nexus: Borders, Labor, and Identity in North America.* Cambridge, MA: MIT Press.

Hekman, R. J. 1984. *Justice for the unborn: Why we have 'legal' abortion and how we can stop it.* CreateSpace Independent Publishing Platform. Available at: https://www.amazon.com/Justice-Unborn-Have-Legal-Abortion/dp/1475124627 [Accessed 23 September 2020].

Keating, A. L. 2012. *Transformation now!: Towards a post-oppositional politics of change.* Ann Arbor, MI: University of Illinois Press.

King, M. L. K. 2010. *Stride toward freedom: The Montgomery story.* New York: Beacon Press.

Kluegel, J. R., Mason, D. S. and Wegener, B., eds. 1995. *Social justice and political change: Public opinion in capitalist and post-communist states.* Piscataway, NJ: Aldine Transactions.

Kooi, A. M. F. 2006. 'Neighbors building neighborhoods: A critical look at citizen participation in Rochester'. Ph.D. Thesis. Cornell University.

Larrain, S., Leroy, J. P. and Nansen, K. 2003. *Citizen contributions to the construction of sustainable societies*. Berlin: Heinrich Böll Foundation.

Lélé, S. M. 1991. 'Sustainable development: A critical review', *World Development*, 19(6): 607–621.

McAdam, D., McCarthy, J. D. and Zald, M. N. 2012. *Comparative perspectives on social movements: Political opportunities, mobilizing structures and cultural framings*. Cambridge: Cambridge University Press.

Mitchell, G. and Dorling, D. 2003. 'An environmental justice analysis of British air quality', *Environment and Planning A: Economy and Space*. 35(5): 909–929.

Mkami, A. 2020. 'Democracy and place in practice: Exploring a community food network'. Ph.D Thesis. University of Connecticut.

Osho. 2000. *Freedom: The courage to be yourself*. New York: St. Martin's Griffin.

Pellow, D. 2018. *What is critical environmental justice?* Medford, MA: Polity Press.

'Principles of Environmental Justice'. 1991. [online]. Available at: https://www.ejnet.org/ej/ principles.html [Accessed September 23, 2020]

Rao, N. and Min, J. 2018. 'Less global inequality can improve climate outcomes', Wiley Interdisciplinary Reviews (*WIREs*). *Climate Change*, pp. 1–6. Wiley Online Library. Available at: https://onlinelibrary.wiley.com/doi/abs/10.1002/wcc.513 [accessed 16 February 2021].

Rose G. 1997. 'Situating knowledges: positionalities, reflexivities, and other tactics', *Progress in Human Geography* 21(3): 305–320.

Schlosberg, D. 2007. *Defining environmental justice: Theories, movements, and nature*. New York: Oxford University Press.

Sen, A. 1990. 'Justice: Means versus freedom', *Philosophy and Public Affairs* 19(2): 111–121.

Sen, A. 2008. 'The idea of justice', *Journal of Human Development* 9(3): 331–342.

Sen, A. 2012. 'Values and justice', *Journal of Economic Methodology* 19(2): 101–108.

Shellenberger, M. and Nordhaus, T. 2005. 'The death of environmentalism: Global warming politics in a post-environmental world'. *The Breakthrough institute* [online]. Available at: https://s3.us-east-2.amazonaws.com/uploads.thebreakthrough.org/legacy/images/ Death_of_Environmentalism.pdf [Accessed September 23, 2020]

Taylor, D. 1997a. 'American environmentalism: The role of race, class and gender in shaping activism 1820–1995', *Race, Gender & Class* 5(1): 16–62.

Taylor, D. 1997b. 'Women of color, environmental justice, and ecofeminism'. In Warren, K., ed. *Ecofeminism: Women, culture, nature*. Bloomington: Indiana University Press, pp. 38–81.

Taylor, D. E. 2000. 'The rise of the environmental justice paradigm', *American Behavioral Scientist* 43(4): 508–577.

The Second National People of Color Environmental Leadership Summit, 2002. Available at: http://old.weact.org/savethedate/2002/2002_Oct_23.html [Accessed 16 February 2021].

The Universal Declaration of Human Rights (UDHR) 1948. The United Nations, pp.art. 21.3.

Torras, M. and Boyce, J. K. 1998. 'Income, inequality, and pollution: A reassessment of the environmental Kuznets curve', *Ecological Economics* 25(2): 147–160.

Warren, K., ed. 1997. *Ecofeminism: Women, culture, nature*. Bloomington: Indiana University Press.

Weber, M. 1958[1905]. *The Protestant ethic and the spirit of capitalism*. New York: Scribner.

Wilkinson, R. and Pickett, K. 2009. *The spirit level: Why greater equality makes societies stronger*. London: Bloomsbury Press.

2 Understanding intersectionality in theory

We have to talk about liberating minds as well as liberating society

Angela Davis

My introduction to intersectionality

It is fitting that the author, political activist, and academic Angela Davis high-lighted the need to "…talk about liberating minds as well as liberating society" (Hutton 2020), given that her book, *Woman Race & Class* (1983) liberated mine and forever changed how I see the world. I first read it in a college women's stud-ies class in 1988 and then went on to use it years later in my classes when I first started teaching. Obviously, when I first read it, I didn't know it would shape what would later become the theoretical focus of my doctoral dissertation and future research, but I did know that her overlapping lenses of gender, race, and class were ones I wanted to continue using. My dissertation, titled "'Sweet Lit-tle Girls'?: Miscegenation, Desegregation and the Defense of Whiteness at Little Rock's Central High, 1957–1959" (Godfrey 2001), built upon Davis' work, looking at the interplay between the socially constructed categories of race, class, and gender, as well as sexuality, in relation to the desegregation of Little Rock's Cen-tral High School. At that time, I was not fully cognizant of how Davis' work itself built upon that of many others, going back to Sojourner Truth's 'Ain't I a Woman' speech at the 1851 Women's Convention in Ohio (Brah and Phoenix 2004), as well as the Combahee River Collective (1977), that existed from 1974–1982 and more. I also was not aware then that such ideas had been further theoretically developed by legal scholar Kimberlé Crenshaw (1989, 1991) into what is now known as *intersectionality*. Although I understood intellectually and theoretically the *need* for an intersectional analysis given the particular historical case study of my dissertation, I was not at the time sufficiently versed in intersectionality's history and its purpose to first and foremost help theorize and achieve *liberation* for African American women. By this, I mean that intersectionality, named by Crenshaw and then further developed by her (Carastathis 2016) as well as subse-quently by other African American feminist scholars such as Patricia Hill Collins (1986, 2000, 2004, 2009, 2011; Collins and Bilge, 2016), is concerned *not only* with theorizing how social identities intersect and transform each other, but more saliently with how they structure society's unequal systems of power on individual, interpersonal, familial, community, national, and global levels, thereby creating a "matrix of domination" (Collins 2000). This matrix is expressed through complex and contradictory relations of privilege and/or oppression that most negatively shape the experiences of low-income African American women, as well as others with highly marginalized intersecting identities such as indigenous people, people deemed to be 'illegal' immigrants, transgender individuals, and/or sex workers. As such, intersectionality from its conception has never just been about merely *theorizing* social inequalities, but rather has always centered progressive personal and social *praxis*, as in Collins' emphasis on "intellectual activism" (2013), in order to achieve authentic social justice that permeates across all of society's intersections. As Collins states, "The framework of race/class/gender studies and its accompanying paradigm of intersectionality …emerged within …[a]… quest for freedom, because it was clear that gaining one kind of 'freedom' (race or gender or class) would not free Black women" (2013, p. 51). In this case, Collins here is talking about the quest for actual *physical* freedom and not just in *theory* – freedom from harm, from neglect, and from inequity – for African American women, and consequently for all oppressed others within society. Additionally, Collins argued that intersectionality necessarily prompts a rethinking of "…the social problems that most affected those most harmed by inequalities—poverty, poor education, substandard healthcare, inadequate housing, and violence …" (Collins 2009, p. viii), with an emphasis on how to better *solve* these problems, not just analyze them. And to do so requires, as she states, "Challenging power structures from the inside, working the cracks within the system… [and thereby]… learning to speak multiple languages of power convincingly" (Collins 2013a, para. 10).

Given that my dissertation was a social history of white racism, as opposed to a form of intellectual activism *addressing* past and current white racism, this original emphasis did not, at the time, resonate with me or with my work. Additionally, given my positionality as a white middle-class female sociology graduate student, who received an egregious lack of graduate-level instruction in Black feminist theory (an academic experience that is sadly not uncommon), I didn't have a "lens", as Crenshaw states, which had "been trained to look at how various forms of discrimination come together" (McCauley 2016, para. 19). Consequently, the activist roots of intersectionality were unknown to me and I did not "develop", building on her full quote, "…a set of …[analytical tools]… as inclusive as they need to be" (McCauley 2016, para. 19). This heavy emphasis on theory, as opposed to the historical roots embedded in social justice practice, is part of a struggle that goes on inside the academy; the radical political intentions of African American feminist scholars are often co-opted, deracinated, and whitened (Bilge 2013; Carbado 2013; Cho et al., 2013), while "intellectual activism" is delegitimized and ironically seen as less rigorous than purely theoretical work lacking practical application (Collins 2013, p. xi). This last point is one I will address more extensively later in terms of my own experiences, but suffice it to say here that I entirely agree with Collins' astute and cutting insight that

...intellectual activists who do devote their attention to the public can pay a high price. In the US scholars and activists who place their education in service to the public are routinely passed over for cushy jobs, fat salaries, and the chance to appear on NPR.

(2013, p. xiv)

In fact, I found her words here quite liberating, as I had come to suspect that the lack of institutional recognition for my work with CLiCK (Commercially Licensed Co-operative Kitchen) was related to my own specific non-tenure-track status at a research-focused institution – reminding me that even sociologists can at times fail to use our sociological imaginations on our own lives! An additional irony of this lack of institutional support for public sociology/intellectual activism is that it was not until I began to work on co-creating CLiCK that I was finally able to fully understand the necessity of intersectionality as a theoretical tool to address the full complexity of attempting to achieve social justice/just sustainabilities (JS) in praxis. This of course links back to my privileged positionality, in that I had never before had to try to create *actual* social change or *actually* put my ideas into practice, as opposed to those whose social oppressions require ongoing attempts to change the/their social reality as part of their daily survival.

Prior to my involvement in this 'public sociology'/'intellectual activism' work, my interests had shifted from desegregation to environmentalism. I wrote a number of articles on ecofeminism and activism (Godfrey 2005, 2008), focusing specifically on the Texan activist Diane Wilson, author of *An Unreasonable Woman: A True Story of Shrimpers, Politicos, Polluters, and the Fight for Seadrift* (2006), and on the Green Sisters of Genesis Farm, an earth-based ecological and spiritual education center in New Jersey. This work enabled me to recognize that under the umbrella of 'the environment', there are places (created by humans or imbued with human meaning) and spaces (physical locations), and these additional and ever-present intersections *also* transform identities and structure relationships of privilege and/or oppression. By this, I mean that I began to see the physical world, including our bodies, as not just the stages upon which unequal social meanings get constructed but rather as co-contributors in both conforming to and resisting those social meanings and realities. Of course, the late nature of this realization is thanks to Western culture's *mistaken* notion of the 'separation' between the social world and the environment (Eisenstein 2019), which induces us to study society as if it is its own entity. This faulty view is not historically shared by the world's indigenous cultures and theorists (Cajete 2000), who, generally speaking, recognize an interconnectedness between the human and the natural worlds. As such, in this culture, we must unlearn our false notion of separation, and so for me, the work of ecofeminism (Diamond and Ornstein 1990; King 1983; Kirk 1997; Taylor 1997a, 1997b), Environmental Justice (EJ) with its 17 principles and JS (see previous chapter), as well as works by indigenous scholars (Cajete 2000; Nelson 2008) have been pivotal.

In 2010, I had the opportunity to be a guest editor for a special edition of the journal *Race, Gender & Class*. I chose to focus on Global Climate Change (GCC), and we ultimately published seven articles from authors around the world on the complex intersections of race, class, and gender in relation to GCC. In the introduction to this volume (Godfrey 2012), I agreed with the view put forth by sociologists Joane Nagel, Thomas Diets, and Jeffery Broadbent (2010) that sociology had been slow to prioritize scholarship on issues related to GCC, despite the field offering obvious tools for analyzing the unequal experiences of GCC in terms of race, class, and gender. These authors noted that the "intersections of race, gender and class" offered "a lens for analyzing *environmental justice* dimensions of global climate change" [italics mine] (Nagel et al 2010, pp. 17–18; also quoted in Godfrey 2012, p. 3). However, as I noted at the time, what struck me as problematic was their marginalizing of such an analysis only to the '*environmental justice* dimension' as opposed to "…the entire issue" of climate change (Godfrey 2012, p. 3). Here, we see again that not only does intersectionality as a theoretical lens get disconnected from its historical commitment to social justice practice, but it often also is narrowed down and only considered applicable when addressing matters related to Black Indigenous People of Color (BIPOC; Bailey et al. 2019) rather than the entire society. Similarly, the terms 'race' or 'gender' or 'class' have come to be used synonymously for those whose racial/gender/class identities bring them oppressions, as opposed to those for whom such identities bring them privileges (Tatum 2000). As a result of this tendency, I sought to make sure that collectively the pieces in this journal edition looked at many different aspects of our intersectional identities. It was at this time that I came across the work of Buddhist monk, teacher, and author Thich Nhat Hanh and his concept of "Interbeing", which I found additionally helpful in developing my understanding of intersectionality in relation to 'the environment'. In fact, I owe my theorizing of intersectionality in relation to the elements of earth, water, air, and fire (Godfrey 2012; Godfrey and Torres 2016, 2016a) to his simple yet profoundly nuanced work. As Thich Nhat Hanh states, "Without a cloud, there will be no rain; without rain, the trees cannot grow; and without trees, we cannot make paper" (Thich Nhat Hanh, 2019, para. 2); all these elements are in essence inseparable; hence they are 'interbeing'. In the introduction to the special edition of *Race, Gender & Class*, I wrote about the connections between 'interbeing' and 'intersectionality', arguing that 'interbeing' "…added an additional dimension in terms of the resulting state, as in our state of being" (p. 6). In other words, interbeing in conjunction with intersectionality captures "… the relational ontological state of 'being'", implying a "…co-constructive, interactive and participatory aspect…". As such, "… that which is being created is not merely the product of an intersection, but rather is itself intersectional, relational and hence inseparable from our 'body subject', hence our 'beings'" (Godfrey 2012, pp. 6–7), which, as discussed, shape our ongoing and ever-dynamic perceptions. CLiCK was created from and out of my/our ideas and theories about what 'could be good', 'what might work'…etc., and with every step, we tried to reflect upon these ideas as we struggled to bring them into being, a struggle that continues to this day. Hence, I came to see CLiCK as a place for me/us to experiment with intersectionality in praxis – which I came to further understand as 'interbeing' – which ultimately led to both CLiCK's ongoing success and to this book, which I am also currently working to bring into being.

After serving as guest editor for this journal project, I embarked on an even more ambitious book project through which I sought to mirror my work with CLiCK in terms of more fully putting intersectionality into being, hence praxis, by delving into its roots and its theoretical uses by other theorists. I wanted to build on my previous work in the journal, but this time not only did I want to more fully link intersectionality with climate change and environmental issues, but I also wanted to edit with others – not just academics but also activists, poets, artists, journalists, playwrights, and dancers from around the world. Collaboration, which is essential for CLiCK and for any social justice project, was a skill I wanted to further develop in relation to writing and editing, although this remains an ongoing challenge. Ultimately, after nearly four years of work, my then co-editor Denise Torres and I published two volumes: *Systemic Crises of Global Climate Change: Intersections of Race, Class and Gender* (2016) and *Emergent Possibilities for Global Sustainability: Intersections of Race, Class and Gender* (2016a), just as Donald Trump was elected president of the United States, an event which made all of the content in both books ever more pertinent.

In our introduction to the first book, we stated that our goal was to "...engage intersectionality heuristically, throughout all aspects of the book" (Godfrey and Torres, 2016, p. 3), both in terms of the diversity of voices included and in the book's organization, focused around the natural elements of Earth, Air, Water, and Fire, as well as Chaos, and in the *Emergent Possibilities* book, we added Aether. In extending the theoretical application of intersectionality to GCC and to all aspects of the environment, we were mindful not to deracinate it in the manner previously discussed or to use it merely as, as we quoted, a "buzzword...reified into a formula merely to be mentioned, being largely stripped of the baggage of concretion, of context and history" (Knapp 2005, p. 255; quoted in Godfrey and Torres 2016, p. 5). In contrast, we sought to

> ...honor intersectionality's lenses as being ground by the struggle in body and mind of a 'broader women's movement where Chicanas and other Latinas, native women and Asian women...[were] at the forefront of raising claims about the interconnectedness of race, class, gender and sexuality in their everyday lived experience'.
>
> (Godfrey and Torres 2016, p. 6 quoting Collins 2011, p. 91)

We sought to preserve this focus by "...including the physical places and spaces their bodies inhabit as salient for analysis" (Godfrey and Torres 2016, p. 6; also see Ducre 2018), examining the ways in which struggle is linked to physical setting. Instead of merely adding space and the environment to an ever-increasing and essentialized list of identities referenced only in passing, we aimed, like Lykke (2010), to reveal the "'human/nature' or 'earth other axis'" (p. 39; quoted in Godfrey and Torres 2016, p. 3), as in being "the foundation component" (p. 3) for any comprehensive intersectional analysis. Hence, by explicitly linking intersectionality to GCC and sustainability, we wanted to call "for matter 'to matter'" (Barad 2008, p. 120; quoted in Godfrey and Torres 2016, p. 4), thereby pushing "the theoretical boundaries" (Carbado 2013, p. 841; quoted in Godfrey and Torres 2016, p. 6), while still recognizing that "…as with any theoretical lens, what is looked at is always partial, situated from a particular social position" (Godfrey and Torres 2016, p. 6). Furthermore, such perspectives are "… not disinterested: they are knowledges made for doing something-truths with a purpose" (Grzanka 2014, p. xxiii; quoted in Godfrey and Torres 2016, p. 6). Our purpose with these volumes was to examine how inequalities are created and perpetuated in relation to the ongoing production of GCC, as well to consider what those inequalities mean for global sustainability, including JS. Now, for this study of CLiCK, my purpose is still to look at the creation and maintenance of inequalities but with particular focus on local and global food systems and a critical lens on this specific example of what can and cannot be collectively done, as well as what has not yet been done.

I often use my co-edited volumes in two of my classes – 'Society and Climate Change' and 'Sustainable Societies' – because they invite students (and readers in general) to see climate change/sustainability through intersectional lenses that focus first and foremost on the highly complex social aspects, which are inseparable from the physical environmental elements. Although these books attempted to theoretically explore what Barad identifies as the 'entanglement' of matter and its ever-evolving social meanings (also see Barad 2007), I have found that CLiCK has challenged me even more in *practice* to understand what such entanglements did, do, and could mean in my/our everyday lives.

> CLiCK has allowed people who were cooking in their homes and selling their food with the dream of someday owning a business to reach that goal. Licensing and other expenses are often too high. Through CLiCK, entrepreneurs have been able to get licensed and have a certified kitchen to cook out of, along with learning about how to better run their business.
>
> CLiCK plants many of its fruits and vegetables. Members of the community are able to learn about farming, healthier eating habits, and safer food preparations.
> Middle-class Latina volunteer

Intersectionality and just sustainabilities

As mentioned in the previous chapter, I was drawn to Agyeman's work because of our shared conviction that if we do not conceptualize the amorphous concept of sustainability as ultimately encompassing social and environmental *justice*, then we cannot even come close to achieving it in the present context, let alone over the 'longue durée'. Yet, as also discussed, justice if not defined or recognized in its full complexity can also evade the totality of itself, even as it may be present for some, in some ways, and at some times. To give a pertinent example, Kimberlé Crenshaw, in her TED Talk "The Urgency of Intersectionality" (2016), asks her audience of about 100 people to stand up and to remain standing until she mentions the name of an African American killed by police in recent years that they have *not* previously heard, at which point they should sit down. She starts by listing the names of such victims as Eric Gardner (September 15, 1970 – July 17, 2014), Mike Brown (May 20, 1996 – August 9, 2014), Tamir Rice (June 25, 2002 – November 23, 2014), and Freddie Gray (August 16, 1989 – April 19, 2015), and slowly members of the audience begin to sit down until by the fourth name only about half are left standing. But then Crenshaw switches from African American male victims to *female* victims also killed by police, women such as Michelle Cusseaux (birthday unpublicized, no Wikipedia page, shot in August 2014), Tanisha Anderson (birthday unpublicized, no Wikipedia page, killed by physical restraint November 13, 2015), Aura Rosser (birthday unpublicized, no Wikipedia page, shot in November 10, 2014), and Meagan Hockaday (birthday unpublicized, shot in March 28, 2015). Nearly all the remaining audience members sit down by the first name, and by the last woman's name only about four are left standing. Crenshaw goes on to say she has done this activity all over the country with women's rights organizations, civil rights organizations, professors (including sociologists), and with progressive members of Congress, and in all cases, the results have been the same: the names of the males are mostly known but those of the females are not, as further evidenced by my internet searches for birthdays. Crenshaw notes that the media has focused on racism against African Americans in general (addressed by #BlackLivesMatter) and sexism against women in general (addressed by #MeToo), but what gets lost are the *intersections* of, in this case, race and gender. To her live audience, Crenshaw declares,

> These women's names slip through our consciousness as there are no frames for us to see them, no frames for us to remember them, no frames for us to hold them… [they] fall through the cracks of our [social justice] movements, left to suffer in virtual isolation. But it doesn't have to be this way!
>
> (2016)

The solution she offers to her audience is of course 'intersectionality', which she asserts can help address "multiple levels of social injustice" (2016) and thereby promote a more complete version, or at least vision, of social *justice*. Crenshaw's audience, who chose to go see her speak and who mostly knew the male names, presumably support racial *justice* in relation to police brutality. At the same time, their ignorance (and perhaps yours, and certainly, to an extent mine) of the female names must nevertheless invite us to question if their (our) awareness of *only* the male names can still count as a demonstration of justice (as discussed in the previous chapter)? If not, then to assure a more complete, more *just* movement, we must engage intersectionality, as Crenshaw argues, both theoretically and in practice, to assure we are not allowing peoples, places, and issues to merely 'slip through our consciousness' and thereby rendering them and their oppressions invisible.

My pairing of JS with intersectionality, in relation to my previous volumes, as well as in my practice with CLiCK and in my current analysis here, is an attempt to try to cover the complexity of the unequal power dynamics throughout society, as well as their corresponding social identities and physical locations. Additionally, it is an effort to ensure that my past and present claims to engage in JS do not come from a partial and thus incomplete frame, whether socially or temporally. JS seek an emphasis on the plurality of experiences, thereby opening up the 'frame' to invite an intersectional analysis, which the activist/scholar/environmentalist Giovanni Di Chiro believes is already there. She argues "many environmental justice activists and scholars…adopt a more relational and embodied approach to environmental politics and pay attention to the everyday" and so tend to be aware of the ways in which "…the interlocking scales of the crisis of climate change creates an intersectional framework for global environmental cooperation that some have referred to as just sustainability" (Di Chiro 2011, p. 234). Overall, I agree with her position that theoretical approaches to the social/environmental nexus, such as some forms of ecofeminism (Mies and Shiva 1993; Taylor 1997a, 1997b; Warren 1997), EJ (as well as the more recent emergence of Critical EJ as spearheaded by Pellow 2018), and now JS, have recognized the ways in which race/class/gender intersect through unequal relations of power, including in relation to place/space. However, I would also argue here that by specifically referencing intersectionality and by linking it to its history, as opposed to the more deracinated 'buzzword' versions that are often used, the emphasis remains on keeping women of color and their liberation at the center of one's theoretical analysis and actual *praxis* (Carastathis 2016). This is not to imply that JS or EJ or even some forms of ecofeminism have not supported such *praxes* but rather that, as Torres and I argued in our second volume *Emergent Possibilities* (2016a),

> …the union of JS and intersectionality theories creates a *means* to doing intersectionality and for evaluating its *ends*: Together they provide a malleable framework capable of linking the parts to the whole with the potential to theorize and put into practice the long-term goals of creating and sustaining a just and livable planet for all beings.
>
> (p. 2)

Given that this is an analysis of what I and others have *done* and have tried to do in the creation of CLiCK, this pairing makes even more sense, for such a framework,

> … provides new tools that are complex enough to analytically encompass the dynamism of both the social and material worlds so that we do not inadvertently reinforce what already exists. Indeed, as a multi-systems framework, intersectionality-informed JS may be used as a personal tool, not just in some distant, hypothetical future, but now, within our individual lives, to evaluate how we interact with all the other lives on this planet, human and non-human.
>
> (p. 3)

This is what I have tried to do in my work with CLiCK: apply intersectionality in the actual pursuit of JS in order to firstly, continually challenge the privilege of my positionality (and that of other white middle-class board members), and secondly, to shift my frames of reference so that the names, both real and symbolic, of our society's most marginalized and oppressed do not 'fall through the cracks'. In this manner, I have come to believe that JS *need* an intersectional analysis to ensure that even when an aspect of justice is achieved, it does not become homogenizing, allowing the possible justice for one to consequently, though *inaccurately*, be seen as constituting *justice for all*. Hence, based on Crenshaw's conviction, I concur that intersectionality encourages, in fact *demands*, ongoing reflective and imperfect scrutiny that endures over time and that seeks to address structural inequalities beginning most specifically with women of color. As Torres and I again stated in our second volume, intersectionality is "not committed to particular 'subjects' nor to identities' but 'to marking and mapping the production and contingency of both" (Carbado 2013, p. 815; quoted in Godfrey and Torres 2016a, p. 3). To illustrate, while Agyeman and colleagues recognize that "the rich can ensure that their children breathe clean air" (2003, p. 1), in our argument "air itself becomes an intersection whose quality – or lack thereof – adds to the creation of how given humans understand themselves and are understood" (Godfrey and Torres 2016, p. 9). As an intersection, we recognize that air (as well as water, soil, food, etc.) changes based on the additional intersections of race, class, and gender identities of the humans consuming it, creating discernable and predictable unequal patterns, as in the disparity of asthma rates with the burden falling on BIPOC, as well as on children and women within these racial/ethnic groups.[1] As such, in this analysis of CLiCK, food in all its complexities is the central focus for my exploration of JS through an intersectional lens. Furthermore, food – as Thich Nhat Hanh (2009) recognizes – is 'interbeing' with water, soil, sunlight, air, pollinators, worms, as well as workers, etc., which all shape the type of food, its quality, and its local and global impacts, as well as whether or not such food can be seen as embodying JS. These points will be developed more in Part II when I focus specifically on CLiCK, although, as will be addressed more in the Conclusion, none will nor can be fully developed, for all such analyses are inherently incomplete, ongoing, and partial no matter how much we seek to see them in the fullness of their complexity.

Can CLiCK launch a family from poverty? I fear it can't achieve that success without other support and focus from other partners – lending programs, ongoing education and technical assistance (financial management, marketing, regulatory compliance, staffing issues), and not pulling out the safety net too early. But it can provide an important piece of the puzzle, one that no one else provides.
Middle-class white female volunteer

Note

1 https://www.aafa.org/asthma-disparities-burden-on-minorities.aspx (para. 1)

References

Agyeman, J. 2003. '"Under-Participation" and ethnocentrism in environmental education research: Developing "Culturally Sensitive Research Approaches"', *Canadian Journal of Environmental Education*, 8: 80–94.

Bailey, J., et al. 2019. 'Getting at equity: Research methods informed by the lessons of intersectionality', *International Journal of Qualitative Methods* 18(1): 1–13.

Bilge, S. 2013. 'Intersectionality undone: Saving intersectionality from feminist intersectionality studies', *Du Bois Review*, 10(2): 405–424.

Barad, K. 2007. *Meeting the universe halfway: Quantum physics and the entanglement of matter and meaning.* Durham, NC & London: Duke University Press.

Barad, K. 2008. Posthumanist performativity: Toward an understanding of how matter comes to matter. In Alaimo, S. and Hekman, S. eds. *Material Feminisms.* Bloomington Indiana University Press, pp. 120–156.

Brah, A. and Phoenix, A. 2004. 'Ain't I a woman? Revisiting intersectionality', *Journal of International Women's Studies* 5(3): 75–86.

Cajete, G. 2000. *Native science: Natural laws of interdependence.* Santa Fe, NM: Clear Light Publishers

Carastathis, A. 2016. *Intersectionality: Origins, contestations, horizons.* Nebraska: University of Nebraska Press.

Carbado, D. 2013. 'Colorblind intersectionality', 38 Signs: Journal of Women in Culture and Society, 4, *UCLA School of Law Research Paper No. 13-19*. Available at SSRN: https://ssrn.com/abstract=2291680

Chiro, G. D. 2011. 'Acting globally: Cultivating a thousand community solutions for climate justice', *Development* 54(2): 232–236.

Cho, S., Crenshaw, K. W. and Leslie, M. 2013. 'Toward a field of intersectionality studies: Theory, applications, and praxis', *Signs: Journal of Women in Culture and Society* 38(4): 785–810.

Collins, P. H. 1986. 'Learning from the outsider within: The sociological significance of black feminist thought', *Social Problems* 33(6): S14–S32.

Collins, P. H. 2000. *Black feminist thought: Knowledge, consciousness and the politics of empowerment.* London: Routledge.

Collins, P. H. 2004. *Black sexual politics: African Americans, gender and the new racism.* London: Routledge.

Collins, P. H. 2009. Foreword: Merging intersections – Building knowledge and transforming institution. In Dill, B. T and Zambrana, R. E., eds. *Emerging Intersections: Race, class, and gender in theory, policy and practice.* Piscataway, NJ: Rutgers University Press, pp. vii–xiv.

Collins, P. H. 2011. 'Piecing together a genealogical puzzle: Intersectionality and American pragmatism,' *European Journal of Pragmatism and American Philosophy* 3(2): 88–112.

Collins, P. H. 2013. *On intellectual activism.* Philadelphia, PA: Temple University Press.

Collins, P. H. 2013a. 'Truth-telling and intellectual activism', [online] *Contexts* 12(1): 36–41. Available at: https://journals.sagepub.com/doi/full/10.1177/1536504213476244 [Accessed September 24, 2020]

Collins, P. H. and Bilge, S. 2016. *Intersectionality. Key concepts series.* Cambridge: Polity Press.

Combahee River Collective. 1977. Combahee River Collective statement [online] *blackpast.org*. Available at: https://americanstudies.yale.edu/sites/default/files/files/ Keyword%20Coalition_Readings.pdf [Accessed September 24, 2020]

Crenshaw, K. 1989. 'Demarginalizing the intersection of race and sex: A black feminist critique of antidiscrimination doctrine, feminist theory and antiracist politics', *University of Chicago Legal Forum*, Article 8, 140: 139–167.

Crenshaw, K. 1991. 'Mapping the margins: Intersectionality, identity politics, and violence against women of color', *Stanford Law Review* 43(6): 1241–1299.

Crenshaw, K. 2016. 'The urgency of intersectionality' [online]. *TED Talks*. Available at: https://www.ted.com/talks/kimberle_crenshaw_the_urgency_of_intersectionality? language=en [Accessed September 24, 2020]

Davis, A. Y. 1983. *Women, race, & class*. New York: Vintage Books.

Diamond, I. and Ornstein G. F. 1990. *Reweaving the world: the emergence of ecofeminism*. San Francisco: Sierra Club Books.

Ducre, K. A. 2018. 'The black feminist spatial imagination and an intersectional environmental justice', *Environmental Sociology* 4 (1): 22–35.

Eisenstein, C. 2019. *Climate: A new story*. Berkeley, CA: North Atlantic Books.

Godfrey, P. 2001. *"Sweet little girls?": Miscegenation, desegregation and the defense of whiteness at little rock's central high, 1957–1959*. Ph.D. Thesis. State University of New York, Binghamton.

Godfrey, P. C. 2005. 'Ecofeminist dialogues: Diane Wilson vs. Union Carbide: Ecofeminism and the elitist charge of "essentialism"', *Capitalism, Nature, Socialism* 16(4): 37–53.

Godfrey, P. C. 2008. 'Ecofeminist cosmology: Genesis farm, ecofeminism and the search for sustainable solutions', *Capitalism, Nature and Socialism* 19(2): 98–112.

Godfrey, P. C. 2012. "Introduction: Race, gender & class and global climate change", *Race, Gender & Class* 19(1–2): 3–11.

Godfrey, P. and Torres, D., eds. 2016. *Systemic crises of global climate change: Intersections of race, class and gender*. London: Routledge.

Godfrey, P. and Torres, D., eds. 2016a. *Emergent possibilities for global sustainability: Intersections of race, class and gender*. London: Routledge.

Grzanka, P. G. ed. 2014. *Intersectionality: A foundations and frontiers reader*. Boulder, CO: Westview Press.

Hutton, B. 2020. "We have to talk about liberating minds": Angela Davis' quotes on freedom [online] AnOther. Available at: https://www.anothermag.com/design-living/12607/ angela-davis-quotes-on-freedom-juneteenth-black-lives-matter-movement [Accessed September 24, 2020]

King, Y. 1983. 'All is connectedness'. In Jones, L., ed. *Keeping the peace: A women's peace handbook 1*. London: Women's Press, pp. 40–63.

Kirk, G. 1997. 'Ecofeminism and environmental justice: Bridges across gender, race, and class', *Frontiers: A Journal of Women Studies* 18(2): 2–20.

Knapp, G-A. 2005. 'Race, class, gender: Reclaiming baggage in fast travelling theories', *European Journal of Women's Studies*, 12(3): 249–265.

Lykke, N. 2010. *Feminist studies: A guide to intersectional theory, methodology and writing*. London: Routledge.

McCauley, M. C. 2016. 'Intersectionality concerns transcend straight, white feminism' [online]. *Baltimore Sun*. Available at: https://www.baltimoresun.com/features/women-to-watch/bal-intersectionality-baltimore-feminism-20160926-story.html [Accessed September 24, 2020]

Mies, M. and Shiva, V. 1993. *Ecofeminism*. London: Zed Books.

Nagel, J., Diets, T. and Broadbent J. 2010. 'Workshop on sociological perspectives on global climate change' [online]. *National Science Foundation*. Available at: https://www.asanet.org/sites/default/files/savvy/research/NSFClimateChangeWorkshop_120109.pdf [Accessed September 24, 2020]

Nelson, M. 2008. *Original instructions: Indigenous teachings for a sustainable future*. Rochester, VT: Bear &Company.

Pellow, D. 2018. *What is critical environmental justice?* Cambridge: Polity Press.

Tatum, B. D. 2000. 'The complexity of identity: "Who am I?"' In Adams, M., et al., eds. *Readings for diversity and social justice: An anthology on racism, sexism, anti-semitism, heterosexism, classism and ableism*. New York: Routledge, pp. 9–14.

Taylor, D. 1997a. 'American environmentalism: The role of race, class and gender in shaping activism 1820–1995'. *Race, Gender & Class* 5(1): 16–62.

Taylor, D. E. 1997b. 'Women of color, environmental justice, and ecofeminism.' In Warren, K. and Erkal, N., eds. *Ecofeminism: Women, culture, nature*. Bloomington: Indiana University Press, pp. 38–81.

Thich Nhat Hanh, 2019. Cultivating insight into interbeing. Available at: https://www.stillwatermpc.org/dharma-topics/cultivating-insight-into-interbeing/ [Accessed 24 September 2020].

Warren, K. 1997. ed. *Ecofeminism: Women, culture, nature*. Bloomington: Indiana University Press.

Wilson, D. 2006. *An unreasonable woman: A true story of shrimpers, politicos, polluters, and the fight for seadrift*. White River Junction, VT: Chelsea Green Publishing.

3   Overlapping lenses

Insights into praxis

> The very nature of materiality is an entanglement. Matter itself is always already open to, or rather entangled with, the "Other".
>
> Karen Barad

Intersecting case studies

Recognizing as Barad does that, "The very nature of materiality is entanglement", it stands to reason that just sustainabilities (JS) and intersectionality, as theoretical lenses seeking to further social justice in practice, would be "...already open to..." each other (Barad 2007, pp. 392–393). To illustrate the benefits of linking them, either overtly or as a consequence of a more nuanced and complex analysis that ultimately seeks progressive social and environmental change, we can look to the work of other researchers and activists, given that my analytical approach is not unique. In fact, as Di Chiro (mentioned in Chapter 2) states, "Many environmental justice activists and scholars adopt a more relational and embodied approach to environmental politics and pay attention to the everyday, material experiences of global environmental problems" (p. 234), as they are on the front lines seeking actual progressive changes in the structures of society. As such, neither Environmental Justice (EJ)/JS nor intersectionality should be seen as ends in and of themselves, nor should their engagements be seen as assurances against the very oppressive structures they seek to overturn. Rather, as Carastathis states at the conclusion of her book *Intersectionality: Origins, Contestations and Horizons* (2016), we should not look to intersectionality "as an epistemological or ethical guarantor, but as a profoundly destabilizing, productively disorienting, provisional concept that disaggregates false unities, undermines false universalisms, and unsettles false entitlements" (p. 237). At the same time, in the everyday, embodied world, such states of being 'profoundly destabilizing, productively disorienting, [and] provisional' can only be sustained in the short term before projects, plans, and actions become overly unsettling, chaotic, and counterproductive. Hence, the key, as will be explored, is to find a balance between the complexities of theory and the embodied demands of praxis, so that one can transparently declare from one's own positionality, that in relation to JS/EJ and intersectionality, "we know them when we *feel* them", even if their destabilizing, disorienting, and provisional qualities make it extremely challenging to both fully articulate them and fully embody them. For to do this requires, as will be further explored, that we recognize as Barad states that they are already "...entangled with, the 'Other'" (Barad 2007, pp. 392–393).

What can help in this balancing project, however, are a few concrete examples, as when Crenshaw asked her audience to sit down if they had never heard the name of a particular African American person killed by the police. Crenshaw did not spend her TED Talk delving into the theoretical weeds of intersectionality, but rather chose to engage her audience in a simple, yet profound, activity that brought the *need* for intersectionality into their very own realities. In this manner, she went from 'theory' to 'praxis', making intersectionality a learning experience in practice and in turn affirming and further reaffirming its theoretical validity.

> *Having a partnership with our local public schools give children the opportunity to learn how to grow their own food, see the benefits of developing skills to put fresh food on their table, enjoy, respect, and honor nature, etc. Another example of promoting social and food justice is our partnership with educators who provide an opportunity for community members to have access to classes that will teach them and make them aware of ways to balance food and social injustice.*
> Middle-class Latino board member

Case studies: intersectionality as praxis

Asian Communities for Reproductive Justice (ACRJ)

This *need* for researchers, activists, and organizations to engage with intersectionality in order to put EJ/JS, or in fact any form of justice, into practice, is supported by an example explored by Di Chiro (2011). Di Chiro seeks to illustrate "...embodied approaches to environmental politics..." that include "...an intersectional framework" (p. 234). She gives the example of the non-profit Asian Communities for Reproductive Justice (ACRJ), based in Oakland, California, which has received an impressive amount of publicity for its progressive policies and practices. In Di Chiro's opinion, ACRJ upholds the "...importance of 'looking both ways', as they put it, through the dual lenses of environmental sustainability and reproductive justice" (p. 235) in order to create innovative strategies that simultaneously mitigate climate change and address the reproductive rights, environmental health, and social reproduction needs of the predominantly low-income Asian and Pacific Island residents of the city of Oakland. ACRJ's approach is articulated in this report, where they state that their "Reproductive Justice frame-work stipulates that reproductive oppression is a result of the intersections of multiple oppressions and is inherently connected to the struggle for social justice and human rights" (ACRJ 2005, p. 1). Their agenda

...places reproductive justice at the center of the most critical social and economic justice issues facing our communities, such as ending violence against women, workers' rights, environmental justice, queer rights, immigrant rights, and educational justice, which have major implications for Asian women.

(p. 7)

The recognition that none of these issues are separate from each other, and that none can be fully addressed unless the others are also addressed, is a central aspect of the EJ/JS and intersectional nexus. Such a nexus pushes our analyses and practices beyond, as Carastathis (2016) recognizes, their superficial "...appropriation by institutionalized mandates of diversity management, commodification, and epistemic totalization" (pp. 238–239). In fact, as Carastathis implies, such appropriation inevitably renders intersectionality as existing merely in the realm of theory, disallowing it to evolve and continually challenge us on "...the horizons of our collective praxis" (pp. 238–239). The same can be said for EJ/JS, in that it is only when evaluated in praxis and over the 'longue durée', not only by outsiders but more saliently by insiders (recognizing that both are needed in the evaluation process), that any such collective assessments of their actual and proposed efficacy can be made.

The ACRJ's report illustrates an example of how their ideological commitments have influenced their practices, referencing their involvement with "The Healthy Communities Campaign, in collaboration with environmental justice groups" (ACRJ 2005, p. 7). This project "...increased the visibility of reproductive health issues related to toxic emissions and culminated in victory when one of the most toxic medical waste incinerators in the nation was forced to close in 2002" (p. 7). Linking EJ issues with those of reproductive health necessitates recognizing the intersections of race, gender, and social class (as well as their structurally negative expressions as in racism, sexism, and economic oppression), given that the women who are most likely to be negatively affected by toxic emissions are poor women and/or women of color, including specifically indigenous women (Hoover et al. 2012). Such linkages when listed may seem to be self-evident, but without an intersectional approach, they easily go unrecognized. To illustrate, on the theme of reproductive health and services, I found a Planned Parenthood report from winter 2008 linking environmental toxins and reproductive health, but nowhere in the report were issues of race, gender, or class mentioned, let alone those of racism, sexism, economic inequality, or sexuality...etc.[1] In fact, the report's front-page image depicted two seemingly healthy and apparently middle-class white women jogging along the ocean, which in many ways overrides the insights of the article, at least from EJ/JS and intersectional perspectives, as such women are statistically the least likely to be environmentally poisoned (Bullard et al. 2007). Hence, not only is such a report offering an incomplete analysis of the environmental factors linked to reproductive health, or the lack thereof, but, more egregiously, it is systematically erasing those most likely to be negatively affected, unlike in the work of ACRJ.

> *Giving opportunity to anyone who has the idea that they want to start a food business – a location, equipment, and support is a major part in our community. The events, classes, workshops, information shared across all populations regardless of income, status, race, etc. put everyone who enters CLiCK (Commercially Licensed Co-operative Kitchen) on an equal level. All people in the community are welcomed and encouraged to learn and explore!*
> Middle-class white female former board member

Deeply intersectional environmental scholarship

Another example of researchers calling for the linking of EJ (not specifically JS in this case) with intersectionality comes from the introduction to a special issue of *Environmental Sociology* titled "Developing deeply intersectional environmental justice scholarship" (2018). At the time of writing, "The EPA's Office of Environmental Justice has been specifically targeted for such drastic cuts and deregulation that long-time head of the division, Mustafa Santiago Ali, resigned in March 2017" (Malin and Ryder 2018, p. 1). To combat such a crisis, the authors assert that environmental researchers such as themselves,

> ...can do much more ... to increase the relevance of our research in identifying core drivers of environmental injustice and then shaping solutions – in part by illuminating the intersections that can at times seem too complex to tease apart or contextualize. This includes continuing to develop and refine exactly how we might more accurately understand, operationalize, and communicate the complexities of these intersections and their manifestation.
>
> (pp. 1–2)

Furthermore, these authors seek to "...demonstrate how environmental sociologists can use deep intersectionality to investigate injustices and thus inform more democratic, multifaceted, and multi-scalar solutions to environmental injustices" (p. 2). For example, one of the articles in this special edition, "The Black feminist spatial imagination and an intersectional environmental justice" (2018) by Kishi Animashaun Ducre, argues that intersectionality has often been lacking in environmental sociology, but has much to offer the discipline; in particular, to Ducre, its saliency is "the inherent spatial nature of this concept" (p. 23). As she goes on to say,

> A visual understanding of spatial intersections and interlocking oppressions connotes an engagement with the distance within and between social identities in the relationship to yet a second spatial dimension: power. The terms lend themselves to a socio-spatial imagining of oppressions that are nonlinear and non-vertical.
>
> (p. 23)

To illustrate, Ducre gives examples from "…community mapping and photovoice projects with mothers on the south side neighborhood of Syracuse, NY…" that revealed, "…distinct spatial strategies of survival and resilience…" (p. 24), meaning that:

> …the mothers did not share the conventional definitions of environment nor did they convey environmental problems as those involving pollution. Instead, abandoned housing, urban decay, and sites of previous or potential violence were recurring themes of environmental problems. Related to this, their limited mobility outside the confines of their neighborhood meant that they engaged in daily strategy of maximizing gains while minimizing their exposure to these environmental risks.
>
> (p. 24)

Recognizing issues of 'urban decay and sites of previous and potential violence' as *environmental* problems and not just social ones, further illustrates the power of intersectionality to bring to light the full intra-acting (Barad 2007) web, wherein the social and the environmental are not separate spheres but rather are theorized, and experienced, as overlapping nodes along intersecting threads, able to equally alter each other's manifestation and articulation. Thus, as Torres and I pointed out in our *Systemic Crises* book, "…doing intersectionality…", which must involve "…examining the whole and its constituent parts simultaneously…", is difficult "…even when there is an explicit commitment to environmental justice" (Godfrey and Torres 2016, p. 9). For example, in the documentary *Fierce Green Fire*, Robert Bullard appeals for the neutrality of air by stating, "There is no Hispanic air. There is no African-American air. There's air!" (Godfrey and Torres, 2016, p. 9; also see Kitchell 2013). However, as we pointed out;

> …although in the abstract, Bullard is correct in that there is only air as a universal, when we locate it materially, in time, place, and history, and examine the situated differences, all air is not equal. In point of fact, from a public health perspective as well as one of environmental justice—as Bullard himself has ironically shown in conceptualizing environmental racism (Bullard, 1993, 1994)—we may in fact experience 'Hispanic,' 'African-American,' or even 'Los Angeles' air, just as internationally daily reports are given on Mexico City and Beijing air, in terms of air quality and the exposure to pollution.
>
> (Godfrey and Torres 2016, p. 9)

Indeed, such intersections of race, class, place (see also McKane et al. 2018), and *air* help illustrate why asthma rates for low-income African Americans and Latinos are higher than for whites, as well as those from the middle class, as Bullard et al. have documented (1994, 2007). Additionally, age is also a factor, with children and the elderly being more vulnerable, and so the intersectional issues become increasingly complex, but also essential for accurate assessment from a public health perspective. However, unsurprisingly, the Centers for Disease Control and Prevention asthma statistics (2016–2018 National Health Interview Survey)[2] list each category separately (age, sex, poverty level, and race), overlooking the ways they intersect and dynamically transform each other from being static to relational.

A key insight from Ducre that emerges from her use of EJ and intersectionality is the recognition that such a synthesis, including physical spaces and places, "… expands the notion of what constitutes an environment to include associated problems of street violence and urban decay, abandoned housing, and neglected and/or failing infrastructure" (Ducre 2018, p. 33). Additionally, as she argues, "…a Black feminist spatial imagination" emerging from such a lens "…casts the women in these neighborhoods not simply as victims, but as resilient agents that thrive and actively engage in placemaking despite multiple challenges" (p. 33). To illustrate, Ducre harks back to a study participant who spoke of how every morning she would wake up to see the "unsightly image" of an abandoned building but that "…she also clung to its potential and reimagined it as a beautiful home for another family" (p. 33). For Ducre, this "…simultaneous orientation of seeing this space as both a burden but a potential benefit is an example of a Black feminist spatial imagination" (p. 33). I also see it as being similar to the kind of imagining that we did throughout the creation of CLiCK (and continue to do) as we take our social 'burdens' and turn them into 'potential benefits' for all, although obviously we have done so from very different positionalities that are mostly privileged.

In concluding this chapter, it is important to remember Broto and Westman's work from Chapter 1 wherein they reviewed 400 flagship initiatives using JS criteria. As noted, they found that many organizations had aspects of one or two of the principles but none had them all. A parallel can therefore be drawn to intersectionality in that so often researchers, as well as the general public, only engage one social variable (like race or gender or class), resulting in an analysis that is not only partial but, in many ways, essentially inaccurate. Likewise, in the case of JS, not only should more of the principles be enacted, but they should also be theorized and analyzed *intersectionally* to ensure that the manner in which the principles are (or are not) being put into practice includes issues of race, gender, and class. Broto and Westman (2017) do not engage or critically analyze race, gender, and class and so it is possible that when evaluating the presence (or lack thereof) of any one of the four principles in relation to the 400 flagship initiatives, they may have missed nuances in terms of how the principles were being engaged and with whom. What may have seemed like commitment to the principle of, "Improving the quality of life and well-being" might have been targeted at a white middle-class community, especially if the initiative did not at the same time meet the third principle of "Enabling justice and equity in terms of recognition, process, procedure, and outcome". Granted, they do emphasize that, "Just sustainabilities is not a ready-made recipe to deliver concrete initiatives, but a set of principles that should guide, rather than dictate, action" (p. 648). Nevertheless, it remains important to emphasize the importance of – no, the essentiality of- overlapping 'thresholds' (Keating 2012), or rather *intersecting* ones, as in JS and intersectionality. This is what I will next attempt to do loosely in relation to

CLiCK, in the hope that we may eventually say with confidence that 'we know them when we *feel* them', however fleeting such a feeling may be. Additionally, what becomes apparent is that although feelings are essential to our measure of our experiences, they are of course highly subjective and as such are not enough. What we need is to aim for internal and external coherence and continuity in theory and in praxis. For as the theoretical physicist and philosopher of mind David Bohm (2002) recognized, if we change our minds to including, "... everything coherently and harmoniously in an overall whole that is undivided, unbroken, and without a border...", then our "...mind[s] will tend to move in a similar way, and from this will flow an orderly action within the whole" (p. xiii). Such an act of creation from 'mind to matter' will be explored in Part II.

If you care about food, eating, and fairness, then you ought to care about CLiCK.
Middle-class white male former Town Manager

Notes

1 https://www.plannedparenthood.org/files/7914/0519/3147/Focus_Winter_2008_web.pdf
2 https://www.cdc.gov/asthma/most_recent_national_asthma_data.htm

References

Asian Communities for Reproductive Justice (ACRJ). 2005. 'A new vision for advancing our movement for reproductive health, reproductive rights and reproductive justice' [online]. Available at: https://www.racialequitytools.org/resourcefiles/ACRJ-A-New-Vision.pdf [Accessed September 24, 2020]

Barad, K. 2007. *Meeting the universe halfway: Quantum physics and the entanglement of matter and meaning.* Durham, NC and London: Duke University Press.

Bohm, D. 2002. *Wholeness and the implicate order.* New York: Routledge.

Broto, V. C. and Westman, L. 2017. 'Just sustainabilities and local action: Evidence from 400 flagship initiatives', *Local Environment* 22(5): 635–650.

Bullard, R.D. 1994 *Unequal protection: Environmental justice and communities of color.* San Francisco, CA: Sierra Club Books.

Bullard, R., Saha, R. and Wright, B. 2007. 'Toxic wastes and race and twenty 1987–2007: Grassroots struggles to dismantle environmental racism in the United States' [online]. *United Church of Christ Justice & Witness Ministries.* Available at: http://www.ejnet.org/ej/twart.pdf [Accessed September 24, 2020]

Carastathis, A. 2016. *Intersectionality: Origins, contestations, horizons.* Nebraska: University of Nebraska Press.

Chiro, G. D. 2011. 'Acting globally: Cultivating a thousand community solutions for climate justice', *Development* 54(2): 232–236.

Ducre, K. A. (2018). 'The Black feminist spatial imagination and an intersectional environmental justice', *Environmental Sociology* 4(1): 22–35.

Godfrey, P. and Torres, D., eds. 2016. *Systemic crises of global climate change: Intersections of race, class and gender.* London: Routledge.

Hoover, E., Cook, K., Plain, R., Sanchez, K., Waghiyi, V., Miller, P., Dufault, R., Sislin, C. and Carpenter, D. O. 2012. Indigenous peoples of North America: Environmental exposures and reproductive justice', *Environmental Health Perspectives* 120(12): 1645–1649.

Keating, A. L. 2012. *Transformation now!: Towards a post-oppositional politics of change.* Ann Arbor, MI: University of Illinois Press.

Kitchell, M. 2013. *Fierce green fire: The battle for a living planet.* Oley, PA: PBS/Bullfrog Films.

Malin, S. A. and Ryder, S. S. 2018. 'Developing deeply intersectional environmental justice scholarship', *Environmental Sociology* 4(1): 1–7.

McKane, R. G., Satcher, L. A., Houston, S. L. and Hess, D. 2018. 'Race, class, and space: An intersectional approach to environmental justice in New York City', *Environmental Sociology* 4(1): 79–92.

Planned Parenthood of Western Washington. 2008. 'The environment and your reproductive health' [online]. *Planned Parenthood.* Available at: https://www.plannedparenthood.org/files/7914/0519/3147/Focus_Winter_2008_web.pdf [Accessed September 24, 2020].

Part II

Understanding from within and without in practice

CLiCK Time Table, 2009–2020

Year	Description
2009	Board Members from Willimantic Food Co-op recognize the need for small farmers to have a place to add value to products. A working group is formed.
2010	Poverty reduction grant awarded – group proposes shared-use kitchen. Year of research confirms need and seeks ways to address issues of urban underemployment and poor health.
2011	CLiCK is incorporated and becomes a 501-(c)(3) non-profit organization based on cooperative values.
2011–2013	Board meets regularly, does more research to affirm and better define need, applies for grants, and searches for a location for the kitchen.
2014	Funds for a building are loaned and a CT DECD (Department of Economic and Community Development) grant for $100k is awarded (funds for building are a match). Also, a $25k DECD loan is acquired and a $25k United States Department of Agriculture (USDA) Rural Business Enterprise Grant is awarded. Purchase of equipment and renovations begin. Another USDA Local Food Promotion grant for $98k is awarded as well as $21k private foundation award to fund the teaching kitchen. A General Manager is hired.
2015	Commercial kitchen opens. Recruitment of members begins. Plans for teaching kitchen are sought. CT Department of Agriculture grant for $43k is awarded. Construction of teaching kitchen completed. Grounds developed to include community vegetable gardens, orchard, beehives, and labyrinth. By November, 14 new businesses have commenced operations; 100 people have attended classes (including summer series for at-risk youth); local organizations including two universities become institutional members; community members become Friends of CLiCK; grant writing and fundraising efforts ongoing.
2016	Teaching kitchen completed (improvements ongoing), classes start, community gardens planted. CLiCK continues to grow with more new businesses (some old, some new) including one whose co-owner became the General Manager. A part-time Processing Outreach and Product Developer is hired, funded by the CT Department of Agriculture. These two additional positions will continue to help CLiCK diversify in the directions of community health and nutrition and processing and product development using locally produced food.
2017	Expansion of business amenities like storage and equipment, development of summer programs.
2018	Member businesses continue to grow but hover at about 20 (fluctuations expected as part of the model), new technology in teaching kitchen, construction underway for family-friendly/accessible bathroom, hoping to further increase capacity/programming in the future. Education Coordinator hired to develop a set of nutrition and culinary classes utilizing the teaching kitchen.
2019	Co-founder and President (me) steps down and a new Executive Director (ED) position is created and funded by a private foundation. New Community Nutrition Educator is hired. Pilot for a Farmer Scholarship program, in partnership with the CT Department of Agriculture, is launched (this has been pretty slow/small so far, but not nothing).
2020	New lower rate Mortgage is secured. USDA Rural Business Development Grant (RBDG) awarded to fund Latinx Business Support Program and staff position of Latinx Outreach and Education Coordinator. Part-time grant writer hired. CLiCK takes on a food rescue role to reduce food waste during the COVID-19 pandemic in collaboration with other local non-profits. CLiCK shifts to a hybrid outdoor/online class model to continue to hold classes while meeting COVID-19 health requirements.

4 Origins of CLiCK from within and without

And suddenly you know: It's time to start something new and trust the magic of beginnings

Meister Eckhart

Introduction: when do beginnings begin? On memories and methods

As I noted in the Introduction, this is my story – rooted in my positionality as well as in my hybrid role as 'the emic and the etic'. Hence, it is based on my recollections, which are interwoven with my notes, Tina's notes, the stories of/from others, survey answers, and collective documents such as CLiCK (Commercially Licensed Co-operative Kitchen) board meeting minutes and other ephemera that form a living mosaic and have collectively created both the physical and the ideological sides of CLiCK. However, although my memories will play a dominant role in this storytelling process, I nevertheless recognize, as Barad (2007) so eloquently states, that,

> Memory does not reside in the folds of individual brains; rather, memory is the enfoldings of space-time-matter written into the universe, or better, the enfolded articulations of the universe in its mattering. ….. Remembering and re-cognizing do not take care of, or satisfy, or in any other way reduce one's responsibilities; rather, like all intra-actions, they extend the entanglements and responsibilities of which one is a part. The past is never finished. It cannot be wrapped up like a package, or a scrapbook, or an acknowledgment; we never leave it and it never leaves us behind.
>
> (p. ix)

As such, my task here in 'remembering' and selecting aspects of CLiCK's story is also a task of 're-cognizing' it, of actively *creating* it, which extends my '…entanglements…' because '…the past is never finished. It cannot be wrapped up…' – even when made into a book. We are, as Janet Fitch says, wedded to "memory" as "the fourth dimension of any landscape",[1] and therefore, it remains manifested individually and collectively, both spoken and unspoken, an ever present, if at times obscured, *intersection*. Yet, the past's obscurity does not make it any less present, nor are we any less 'responsible' for the ways and means of telling and receiving its stories and making them manifest in the present. In fact, as Thomas King (2008) astutely recognizes, we must "be careful" with the stories we tell (p. 10), as there is great power in the words and meanings they convey, for it is through them that we humans construct our social and consequently physical worlds (Berger and Luckman 1967). In fact, in further building on Barad's (2007) work wherein she seeks to recognize the workings of the universe as being made up of dynamic "entanglements", it is impossible for me to identify a single starting point for CLiCK, for emergence does not happen in any singular fashion. Rather, as Barad (2007) observes, all "…matter and meaning, come into existence, are iteratively reconfigured through each intra-action, thereby making it impossible to differentiate in any absolute sense between creation and renewal, beginning and returning, continuity and discontinuity, here and there, past and future" (p. ix). In other words, I cannot say exactly when and how CLiCK as an idea, as an invention, and as an articulation began, who specifically began it, and how such a complex and 'entangled' process actually got started, for ultimately, as Barad (2007) and Butler (1999) recognize, identities, like all matter, are constituted together.

That said, I will simplify CLiCK's origin story out of necessity, asking the reader to trust that I do so not to deceive or advance my own agenda (apart from, of course, my 'agenda' for larger social justice benefits), but rather because I must take what is an 'entangled' entity, a mix of autoethnography, collaborative ethnography, and more standard ethnography, and turn it into a comprehensive coherent narrative that fills up these pages. Hence, one of my research methodologies is the act of remembering, of untangling, of attempting to speak of singular, isolated events, of beginnings and endings, or of, as T. S. Eliot wrote in his poem *The Love Song of J. Alfred Prufrock*, "…decisions and revisions which a minute will reverse" (Eliot 1917, p. 12). Additionally, given the overall theoretical lenses of just sustainabilities (JS) and intersectionality, I will also engage these frameworks methodologically in order to further dissolve the illusion of separation between the researcher, myself, and the object of my research, as well as "…the imagined duality between truth and falsehood" (Godfrey and Torres 2020, p. 284), in that their categorizations are situationally, contextually, and *intersectionally* dependent. Of course, as I hold the position of both the emic and the etic, such dissolution is inevitable. Even the parts of this story that are not directly about me can only be recounted through placing my positionality on the table, as opposed to claiming to be exercising the ideal of scientific objectivity. As Barad (2007) has demonstrated with quantum physics, such claims of objectivity are fictitious, and yet, this does not release us from accountability; in fact, Barad emphasizes that such entanglements 'extend' our 'responsibilities'. Furthermore, as the theorist Chela Sandoval (2000) asserts in her work *Methodologies of the Oppressed*, the way research is constructed and enacted dramatically changes both what is observed, as well as the outcomes of such observations. Hence, for a methodology to address social injustices, it must be grounded in "'love'" and "…understood as a technology for social transformation" (p. 2), not just for its own sake. Lastly, what begins as 'my story' will eventually bifurcate into also being CLiCK's story, hence a collective story, which although told here from my vantage point is, in the end, a separate entity.

In terms of framing JS as a methodology, Agyeman has previously recognized, in relation to environmental education, that "…culture" is "…central to the research process" (Agyeman 2003, p. 83), influencing how knowledge is constructed, as well as whose knowledge even gets recognized as such. This includes the culture position of 'the other' as well as that of the researcher, who has more often than not been an ethnocentric white male from the Global North. This point is especially salient if the *other* has previously been silenced, objectified, or misconstrued by such researchers, as has been and still is so much the case as a result of racism, sexism, classism, colonialism, and all other forms of dehumanization within the academy, despite frequent claims of objective and politically impartial research. In fact, claims to 'objectivity' and 'impartiality' that obscure the researcher's positionality contrastingly leads to bias and even oppressive research, as opposed to when such claims are not made, as in my case, or that of others mentioned. This is because ultimately, we recognize that 'objectivity' and 'impartiality' cannot ever be fully achieved (Madison 2012). Hence, disclosure of the researcher's intersectional positionality allows for transparency by recognizing that the observer and the observed are not ultimately separate.

As for intersectionality as a methodology, according to Christensen and Jensen (2012), there has been a lack of "…debate about concrete intersectional methodology and analysis" (p. 110). However, given intersectionality's radical roots, they emphasize that "…the method of practicing intersectionality must be related to power relations, in particular locations and contexts…" while still recognizing that "… there is more than one way of doing intersectional analysis" (p. 121). The inherent variety of this approach is clear, as "… in any specific analysis it is necessary to select a number of categories or establish anchor points as a strategic choice" in order to make the analysis possible and enable researchers "…to focus on the categories that are deemed most important for a specific research question at a specific time" (p. 112). One such approach is a focus on "life-stories", which "…can be considered an important method in intersectionality research" in that "…narratives tell us how people draw on different categories in the construction of their life-story" (p. 114), as already illustrated in my own positionality. Of course, our life-stories always obscure some identities, while highlighting others, depending on numerous other influences. Thus, Christensen and Jensen aptly recognize that although "…life-story narratives do contain important actual information…it is also important to stress that such narratives can be analyzed as representations", which "contain information about subjectivity and collective processes as well as social structures and institutions" (p. 14). To give an example, it is common for whites to not speak of their white privilege in their life-stories, but this does not mean the privilege is absent (Tatum 2000). Therefore, the role of the researcher is to link the individual's story into a larger social context, just as I am doing here with my story in relation to CLiCK and through my engagement with Weber's concept of 'ideal types', as in 'ideal type individuals' (Swedberg 2018) used to illustrate what our social identities and positionalities mean in relation to our actions.

> *I do think CLiCK helps to promote sustainability by facilitating businesses that can then buy ingredients from local farms (like Azuluna) or grow ingredients right in the garden (like Dragon's Blood), which all helps to support the resiliency of the local food system and reduces the amount of transport for food products. I think a shortcoming is the lack of a composting program, and the ongoing difficulties with getting actual farmers to use the kitchen themselves.*
> Middle-class white female board member

CLiCK micro origins: for places-spaces and narratives

In speaking of CLiCK's origins, I posed above the philosophical question, as well as the practical difficulty of identifying when and how it began, but I failed to engage with the question of location, as CLiCK is located on a physical site and so it seems self-evident. However, I want to re-emphasize *critical* EJ scholar David Pellow's (2016) argument that in relation to place we need to recognize "…multiple scales, from the cellular and bodily level to the global level and back" (p. 4), and in so doing, 'scale' becomes an *intersection*. Additionally, the geographer Doreen Massey (2005) recognizes the ongoing interplay between place (assumed to be humanly created, meaningful, and thereby knowable) and space (assumed to be the *a priori* matter beyond, unknowable, that is then turned into place). Massey argues for the recognition of space as the amorphous stuff from which places are constituted and as "…a product of relations-between, relations which are necessarily embedded material practices which have to be carried out, it is always in the process of being made. It is never finished, never closed". In fact, she calls for space to be imagined "…as simultaneity of stories-so far" (p. 9) and argues that it is only in recognizing space as such that there exist the possibilities for "…multiplicity and heterogeneity…" (p. 11) and ultimately for space and the future to be open to being formed, hence able to be imagined (p. 12). Thus, when I refer to CLiCK, although I do think of it as a 'place', a building on a parcel of land, I want to also imagine it '…as simultaneity of stories-so far', as unfinished (like the past is for Barad) and open to being reconstituted, both in memory and in collective reality, as opposed to being fixed, closed, and singularly construed. In this manner, I also want to again recognize that the land CLiCK currently inhabits was once Indigenous land, of the Mohegan-Pequot Nations (the boundaries overlap[2]) as mentioned in relation to our interfaith land blessing led by a Mohegan elder (see Preface and Figure 4.1) which included prayers from all the Abrahamic faiths, as well as Buddhism, Hinduism, and Paganism (Figure 4.1).

And so, despite the ongoing legacy of conquest and genocide, since the past is never finished, the present story of this land continues, emerging from its

Figure 4.1 Interfaith Land Blessing Led by Chris 'Painted Turtle' Harris, Mohegan Tribal
   Elder.

diverging roots yet, we believe, still able to transubstantiate its energies into new
positive and healing manifestations.

The difficulty in pinpointing 'beginnings' comes from the fact that, like all
human endeavors, CLiCK began as an amorphous, collective idea, unbound in
its nascent stages by the limitations of reality. As such, I could easily argue that
CLiCK's roots lie in my mother's social service work, or in my childhood desire to
be a chef/baker, or in my years growing up in Europe where food is as sensual as
it is sacred, or in my work as a bread baker during college for a small natural food
store in Princeton, New Jersey, or in my identification as a socialist in college,
leading to my yearlong involvement at Rutgers University Anti-Apartheid Divest-
ment Protest (1985),[3] or my years of work with and for the rights of the develop-
mentally disabled, or my service as a Peace Corps volunteer from 1989 to 1991 in
Cameroon, West Africa, or the development of my first Sociology of Food course
in 2002, or in my idea to start a student-run farm at University of Connecticut
(UCONN) that became Spring Valley Student Farm,[4] or in my involvement with
an anarchist collective in Willimantic called The Wrench in the Works (that no
longer exists), or in my meeting of Tina in 2008 that led to an 'Imagine Willimantic' project, which then overlapped with our involvement with the Willimantic

Food Co-op, which then led to the vision for CLiCK. However, in a given narrative, we must begin somewhere and from there weave the threads of memory using the colors we deem most salient, and so, I will begin with the Willimantic Food Co-op.[5] Additionally, storytelling requires characters whose actions, along with my interactions *with* them and my memories of those actions, drive the narrative and give it flesh. However, as mentioned, although such characters are based on real individuals, they must also be seen as 'ideal types', acting within their socially prescribed roles based on the intersections of race, social class and gender, as well as their professional roles, as in a 'white female grocery store manager' 'a white male philosophy professor' 'a white female small-scale farmer' …etc. These intersecting social roles influence and can even dictate an individual's behavior, as so starkly demonstrated by Philip Zimbardo's 1971 Stanford Prison experiment.

In this famous experiment, self-selected male and mostly white students – who were presumably attracted to a prison experiment in the first place – were put into highly scripted roles of 'guards and prisoners', thereby playing out the socially identifiable roles of *actual* 'guards and prisoners'. Writing in *The New Yorker* about this experiment and another one done in the United Kingdom in 2007, Maria Konnikova (2015) concludes that what these two experiments teach us is not so much our propensity "…for tyranny or victimhood…" as proposed, but rather that "…our behavior largely conforms to our preconceived expectations. All else being equal, we act as we think we're expected to act—especially if that expectation comes from above" (para. 17). Hence, putting deeper philosophical questions of 'free will' aside, it is essential to note that social identities emerge from behaviors that become recognizable within existing social categories and in so doing simultaneously recreate those very same categories. Henceforth, when I mention an individual in a role, including myself, it should be kept in mind that we are in these roles because we are adequately performing them and that the one who 'performs' and the 'role' that is played are not separate but are co-created. This idea supports Barad's notion of entanglements, as well as Judith Butler's concept of 'gender performativity', which as she states in her seminal book *Gender Trouble* (1999) "…constitut[es] the identity it is purported to be" (p. 33). This concept can also be applied to our other socially prescribed roles, which rather than expressing some inner aspect of our individual identities are, as Butler again says about gender, "always a doing, though not a doing by a subject who might be said to preexist the deed" (p. 33).

The amorphous ideas that became CLiCK stemmed from the Co-op's collective desire to seek achievable solutions for some of the ubiquitous problems within the industrial food system, namely food waste and lack of economic returns for small farmers in eastern CT. Of course, these problems, among so many others in relation to the industrial food system, are global in nature and are directly linked to the *inherent* dysfunction of capitalism in relation to the food system. In fact, as Eric Holt-Gimenez argues in *A Foodie's Guide to Capitalism* (2017), "To call the system broken is to believe it once worked well for people, the economy and the environment… The food system is not broken; rather, it is working precisely as a capitalist food system is supposed to work" (p. 56). In other words, the problems of the current capitalist food system are 'by design'. Therefore, attempts to address these 'by design' problems both in our community and globally must work outside the rules of capitalism, such as the sacred tenet of 'private property', by ensuring cooperative principles, including shared access to capital. In CLiCK's case, this understanding ultimately led to our creation of shared-use commercial and teaching kitchens.

Collective ideas do not have singular locations in individual minds but rather act as "emergent properties" (Capra 1997) or "emergent strategies" (brown 2017) within specific social contexts, giving evidence to the adage of being 'more than the sum of their parts'. In this case, the specific social context from which the idea emerged was the Co-op's board meetings in the spring of 2009, when I was the board president and Tina was an officer. Tina had worked for the Co-op for a number of years and had been involved socially for even longer, and so when we got together, we felt drawn to serve on the board. When I first moved to Willimantic in 2005 to take a two-year teaching position at Eastern Connecticut State University (ECSU), I was happy to learn of the Co-op's existence and to read in their brochure about the history of co-ops going back to the early 19th century in Rochdale England (new evidence has also found examples from 1761 in Fenwick, Scotland[6]) by industrial weavers known as the Rochdale Pioneers (Fairbairn 1994). Given my affinity for social justice movements, and the influence of my parents, who were from England and were also supporters of alternative ideas, places like co-ops always felt like home to me. Tina felt similarly for her own reasons, linked to identifying with late 1960s hippie culture. Hence, serving on the board made sense to us individually and also helped to secure our emerging love relationship into a larger social mission (more on this linking of love and labor later).

The Willimantic Co-op began as a buyer's club in the late 1970s, became an incorporated entity in 1980, and has since grown over the last 40 years into a vibrant store located on Prospect Street in Willimantic, CT, with an annual revenue of about $2.1 million and about 16 full-time workers. As the state's oldest co-op, the Co-op is, as stated on their web site, "...a member owned and operated food store"[7] that supports the seven cooperative principles,[8] including a commitment made early on to be located in Willimantic, as opposed to the neighboring town of Mansfield, CT, that is much whiter and wealthier.[9] The Co-op is legally a B-Corp, not an official co-op, as neither the workers nor the members own the profits. However, the Co-op does give dues paying members (the majority of whom are white, socially progressive, largely college-educated, middle-class) a certain amount of power over decisions (such as a recent question about expanding/moving/building a kitchen) through voting and a lower price on items than non-members.

It was out of this white, socially progressive, largely college-educated, middle-class entity, represented by its board of directors (of which we were part) and manager, that the collective vision of creating a shared-use kitchen where local farmers (most of whom also fit this demographic and often came to farming as a social choice rather than as a family heritage) could add value to their products, thereby increasing their farm revenues while also reducing food waste, emerged in 2009. So compelling was this vision of a co-op processing kitchen that a new subcommittee was formed including a number of non-board members, among them a social worker/community organizer from Hartford who had been working in Willimantic to promote community engagement around ideas for economic development. Echoing the words of Meister Eckhart, the Christian mystic, collectively it seemed "...time to start something new and [to] trust the magic of beginnings" and thus this was the micro origin of CLiCK. Furthermore, in understanding the fecundity of the term 'magic', I appeal to the work of David Abrams (1996), in his lyrical book *The Spell of the Sensuous: Perception and Language in a More-Than-Human World*, to not only see magic as is currently popular, as in "... the ability or power to alter one's consciousness at will" (p. 9). Rather, my experiences with CLiCK have helped me affirm what Abrams identifies as,

> ...its ... most primordial sense ... the experience of existing in a world made up of multiple intelligences, from the swallow swooping overhead to the fly on a blade of grass, and indeed the blade of grass itself-is an experiencing form, an entity with its own predilections and sensations, albeit sensations that are very different from our own.
>
> (pp. 9–10)

As such, magic is better understood here as our reawakening to the complexity and interconnectedness of ourselves, the world, and therefore of everything, as will be further explored.

What excited everyone about this new idea was the realization that kitchens bring food to life in ways that store shelves and freezers do not. The notion, as well as the fantasy (as will be discussed more later) that locally grown food could be locally processed and given added value through extended shelf life added a hitherto missing production piece to the Co-op, completing a link on the local/regional food chain. Early meetings were therefore well attended at specific board members' houses, including our own, and conversations flowed as the ideas and questions emerged such as: "What could be seasonally processed?"; "How could the Co-op acquire more space to add a kitchen?"; "How could grab-and-go options such as soups, salads, and sandwiches be included?"; and "How would such a kitchen help further the Co-op's commitment to local farmers, to sustainability, and to a more just and equitable food system?".

Yet, as all activists and social change agents know, talk is easy, talk is cheap, especially if those doing the talking have their most immediate social/physical needs already met, making the visions and ephemeral ideas of 'wouldn't it be nice' lack urgency. And so, these original Co-op kitchen meetings continued for several months without concrete plans or actions. At the same time, tensions were arising on the board among several members, including myself, Tina, and a few other officers, surrounding the then-manager who was controversially also a voting board member (a recognizable conflict of interest). In fact, as board president, I had been approached by a number of the Co-op workers asking for the board to address grievances with the manager, even though that was not within the Board's purview, as well as ideas for unionization. I remember being struck by the realization that the Co-op might not be as equitable or as desirable a place for some of its workers as I had previously imagined based on its overall social justice ethos. Additionally, I realized that addressing this potential contradiction between the organizational image (Terkla and Pagano 1993) of the Co-op as representing social justice within the community and the apparent reality for some (not all) of its workers would be highly challenging; institutions, like individuals, seek to preserve themselves and their images when under threat (Currie et al. 2012; Lawrence et al. 2009). However, navigating such contradictions and conflicts would become par for the course, as Tina and I sought to put forth our vision of what we thought justice should look and feel like, especially at a co-op with its seven cooperative principles, resulting in her quitting the board and me being inadvertently removed.

At this same time, I was contacted by one of the original Co-op's founders, who was a retired white male UCONN philosophy professor and who was working with a local social service agency to see if the cooperative model could help to address issues of poverty and inequality in Willimantic (more on Willimantic's demographics in the next section). The social service agency had received in November 2009 a "Break out of Poverty" Poverty Reduction Incubator Grant for $100,000, funded through the American Recovery and Reinvestment Act (ARRA) of 2009.[10] The plan for the grant was to parcel it out in sub-grants of $10,000 to ten poverty reduction proposals; awardees would then spend a year developing a "...business plan demonstrating how the selected initiatives could link to and leverage local resources by working with non-profit agencies, state & federal agencies and private industries to meet the job development criteria" (personal communication, April, 2010). The only catch, I was informed by the retired professor, was that applications were due the next day. I therefore took the initiative and used the ideas already developed by the Co-op kitchen subcommittee, adding a poverty reduction aspect by arguing that the processing of locally grown fruits and vegetables could be done by those needing job skills and/or seeking to get back into paid employment. I also envisioned that the kitchen could have health and nutritional education, as well as culinary job training, and could link the mutual economic needs of both the urban unemployed and underemployed with those of the small regional farmers we had originally identified. Additionally, such a model could further extend cooperative principles by bridging the racial and social class divides so apparent in both the town and the Co-op. I named the proposed project CLiCK as an acronym for 'Commercially Licensed Co-operative Kitchen', thereby officially marking it as a separate entity (previous names included the Willimantic Co-op Commercial Kitchen and the Cooperative Kitchen Initiative) and submitted it by the due date.

We were awarded the $10,000 grant, thereby laying the groundwork to turn CLiCK from an idea into a potential viable reality. A similar idea had been submitted by the local soup kitchen[11] director who felt that those served at his kitchen would benefit from learning culinary job skills, while bridging the social divides between those who are served (the homeless and those lacking in food security) and those who do the serving. His proposal was not funded and so we combined our ideas to further ground CLiCK into a social justice approach.

Then, just when I thought interest from the Co-op and initial subcommittee would become increasingly invested, as funding can have a solidifying effect in terms of concretizing ideas into material reality, the opposite happened. Tina and I ended up being the only ongoing members, as well as a social worker from Hartford, and so we began recruiting others to help with implementing the grant and the new vision of linking farmer economic needs with the needs of those who are unemployed and underemployed. Hence, as a result of the focus of this grant, CLiCK became committed, at least on paper, to issues of social justice and economic equity.

Now, if this were a non-sociological telling of the CLiCK story, I could leave the drop-off in interest from the Co-op as merely coincidental or I could light-heartedly speculate, as Tina did at the time, that though many had once been excited by the kitchen idea, the acquired funding now meant action and account-ability as opposed to just *talking*. Or I could blame myself, assuming that perhaps other members felt slighted by the last-minute decision to apply, as I had not been able to secure the group's approval (except for Tina's) ahead of time, even though I stuck to our original commitment to local farmers, while adding the link to poverty reduction. However, from a JS and intersectional lens, I cannot ignore the significance of what occurred after we were awarded the grant from the perspectives of race and class. In fact, the loss of participation from the original Co-op members, including the once highly committed manager, speaks directly to the demographic and ideological differences between the Alternative Food Movement (AFM) and the Food Justice Movement (FJM), as mentioned in the Introduction. For, as Zitcer (2017) argues, although food co-ops can offer an alter-native economic model, doing so successfully while also "…living up to coopera-tive ideals is difficult…", especially given their tendency to be "…inaccessible in term of prices, selection and cultural relevance" (p. 182). To reiterate, the AFM, a largely white middle-class movement (e.g., Alkon and Agyeman 2011; Bradley and Herrera 2016; Guthman 2007; Slocum 2007) wherein supporters seek better food and better opportunities for small-scale white farmers, is distinct from the FJM that has been mostly spearheaded by African Americans and Latinos, who also seek better food while inseparably addressing the structural inequalities based on race, social class, and gender that make it difficult to afford/access healthy food or grow it without access to land and/or resources (Alkon and Agyeman 2011; Guthman 2007; White 2018). Therefore, ironically (or not), in combining the interests of those from the Co-op committed to the AFM with the racial and equity issues addressed by the FJM, I had crossed an ideological line that, like the complaints from the Co-op workers, revealed divisions of both social class (Muzika, et al., 2019) and white racial fragility (DiAngelo, 2011, 2018; Fannon 2008). This 'crossing', along with the increasing tensions Tina and I were expe-riencing with the Co-op manager, not only ended our time on the Co-op board as mentioned, but also negatively shaped all future Co-op/CLiCK relationships (these are finally improving, I believe, now that we are no longer involved) in ways that will be further explored. Suffice it to say here that the severity of the resulting negative repercussions ironically revealed to me not only the realities of the feminist adage that the 'personal is political', but more significantly that interpersonal micro-conflicts can and often do undermine the larger more seductive trappings of progressive political commitments. As a result, they make any attempts to expose apparent contradictions and hypocrisies between external 'social justice' claims and internal oppressive practices extremely socially and emotionally challenging, as will be further explored.

> *There is a great concentration of creative energy in the shared kitchen. While individual members are pursuing their own goals, there is a constant sharing of thoughts, methods, possibilities, and plans, resulting in a very dynamic atmosphere.*
> Middle-class white male kitchen and board member

CLiCK macro origins

As mentioned in the Introduction, the AFM can be seen as beginning in the 1970s, when seminal writers, such as Frances Moore Lappé (1971) and Wendell Berry (1977), as well as more recently Michael Pollan (2006, 2013), introduced on a national level the negative impacts of the industrial food system (including its emphasis on meat eating) on animals, humans, and the planet. The AFM was a response to the ever-increasing industrialization of food production, characterized by mechanized farm operations, corporate agriculture monopolies, high usage of chemical fertilizers and antibiotics, as well as negative health impacts on farm workers, farm animals, environmental systems (like soil, water, and air pollution), and consumers (Alkon 2012; Alkon and Agyeman 2011; Bell 2004; Horrigan et al. 2002; Magdoff et al. 2000; Patel 2008; Shiva 2000). The AFM sought solutions embedded in organic farming, vegetarian diets, community gardens, farmers markets, and cooperatives (Berry 1977; Lappe 1971; Pollan 2006, 2013), among other initially small-scale alternatives (like the Willimantic Co-op). Such alternatives have, over the years, resulted in tremendous economic growth of the organic food market, epitomized by such successes as Whole Foods, as well as the availability of organic food at Walmart, which has the largest share of grocery sales in the country (Chait 2019). However, even though AFM advocates seek to recognize, as the African American food justice guerrilla gardener Ron Finley states in his 2013 TED Talk, that "Food is the problem and food is the solution" (Finley 2013), they nevertheless for the most part overlook (from a combination of ignorance, discomfort, racism, and classism) the aspects of the 'food problems and food solutions' that stem from and are inseparable from issues of race and class, in particular when it comes to white middle-class privilege. I see these same tendencies in CLiCK's initial origins, as originally the idea for a commercial kitchen was well supported by the Co-op board and manager until I wrote the aforementioned grant and connected larger social dots that were already there, but which we had not seen or had chosen not to see based on our homogeneous white middle-class positionalities. This connection of the interests of the more individualized AFM with those of the more collective FJM seems to have contributed to the withdrawal of the previously interested parties from the project, in my current sociological understanding.

Adding to this conclusion is the fact that the AFM, epitomized by the notable authors previously mentioned, has overall made little to no reference to the historical and more recent food and agriculture experiences of Black Indigenous People of Color (BIPOC) and/or those for whom food is not embedded in the luxury of individual choice but rather in the stressors of daily survival. As Alkon and Agyeman (2011) note, the AFM has been "...something of a monoculture" (p. 2) wherein "...food movements adherents must be willing, and also able, to participate by *purchasing*" [italics added by author] (p. 3) their *alternative* food choices. This recognition that the AFM is in fact a movement based mostly on one's *individual* economic participation as a consumer, as opposed to one's *collective* role as a citizen or as a member of a specific local community, with a specific cultural history, as well as with a specific place-based ecology and food system, potentially including 'the commons' (Godfrey and Freake 2016), is essential to further distinguish it from the FJM. This recognition of the differences in terms of 'individualism versus collectivism' speaks to the ways and means that 'individualism' is a product of intersecting privileges, allowing middle-class whites to engage in the AFM as individual consumers making alternative food choices, as opposed to being part of a collective struggle for food justice, as with the FJM.

To further illustrate, Monica White (2018), in her groundbreaking book *Freedom Farmer: Agricultural Resistance and the Black Freedom Movement*, recognizes one of the three strategies that Black farmers, both historically and currently (as in the FJM), have engaged in *resistance* to racism and other forms of social, hence food, oppressions, has been by formulating what she identifies as, "...commons as praxis". The other two forms of resistance she identifies are, "...prefigurative politics" that she defines as "...place-based alternative practices, and experiments in everyday living..." and "economic autonomy" that "...allowed communities to provide for their members financially and to help them move from dependence to independence" (p. 10). And despite obvious and seminal historical and contextual differences, these three forms of resistance nevertheless define aspects of CLiCK in theory and in praxis. In fact, "commons as praxis" is the most salient of White's concepts and so I quote in full:

> Commons as praxis engages and contests dominant practices of ownership, consumerism, and individualism and replaces them with shared social status and shared identities of race and class [and, I would add, gender]. It functions as an organizing strategy that emphasizes community well-being and wellness for the benefit of all. It is based on the premise that pooling resources can transcend the limitations of individual strength in oppressed communities. It emphasizes that shared ideology and cooperative/collective behaviors that

arise in response to the conditions of oppression. Community decisions made around shared space and resources such as access to land, and seed [and, I would add, commercial kitchen spaces…] are examples of commons as praxis.

(pp. 8–9)

In linking White's insights of Black farmer resistance with my analysis of CLiCK, I do not seek to make them seem the same by erasing the stark differences, as well as those that are more nuanced. Rather, I do seek to affirm that any solutions to our racist, oppressive, and inherently unequal society, and hence food system, must be cooperative/collective, democratic, and 'emphasize community well-being and wellness for the benefit of all', and that as such CLiCK embodies aspects of 'commons as praxis'.

Given the long history of resistance on the part of Black farmers by creating cooperative farms, such as the Freedom Farms Cooperative (FFC) as detailed by White (2018), the FJM is characterized by the recognition that "…the food system itself …is a racial project …[that] problematizes the influence of race and class on the production, distribution, and consumption of food" (Alkon and Agyeman 2011, p. 5). In addition, like the AFM, the FJM seeks to create and support "…environmentally sustainable alternatives"; however, the FJM, as discussed, explicitly seeks *collective* alternatives that will "… provide economic empowerment and access to environmental benefits in marginalized communities" (p. 6), not just primarily for individuals from communities of privilege. Deeply rooted in the EJ and JS movements (as discussed in Chapter 1) and their recognition of the institutional levels of structural inequalities based on race, class, and gender, the FJM specifically focuses on "…the concepts of *food access* and *food sovereignty*" [italics in original] (p. 6), with food access understood as access to culturally appropriate, affordable nutritious food and food sovereignty understood as the right to define and control one's own food and agriculture systems, as epitomized by the work of the international peasants organization Via Campesina.[12] Additionally, recognizing that food is culturally and physically intimate (Winson) and yet ever fluid and evolving, the FJM seeks to embrace these more personal and communal aspects of food, while nevertheless still highlighting the racist and oppressive aspects. These oppressive aspects result from the legacies of colonialism and slavery (Holt-Gimenez and Harper 2016), emerging into government and corporate policies, including the Farm Bill (Elsheikh 2016), resulting in "…an environmentally and socially destructive centralized agribusiness system" (p. 12), within the larger global capitalist system, under which all our food choices (and arguably most of our other choices) are made.

In returning to CLiCK's poverty reduction grant from a macro origins perspective, according to my email record, the grant "…was designed to nurture new, innovative, sustainable and high-impact replicable poverty reduction programs designed to increase the capacity of low-income and/or no-income people to break the cycle of poverty by increasing their ability to earn income and accumulate assets" (personal communication, May, 2010). Given that the Town of Windham (of which Willimantic is a census-designated area) was ranked by a 2012

Community Food Security assessment as 165th out of 169 CT towns in terms of population at-risk of being food insecure (Rabinowitz and Martins 2012), such a poverty reduction incubator grant program was merited. Once a cotton and silk production center with six mills owned by the American Thread Company (CT's largest manufacturer in 1880)[13] – including, it is said, the first factory in the USA to get a second shift with Edison's newly invented light bulbs (Nunez 2014, p. 35) – Windham and many of its citizens plunged into poverty when the company closed up shop in 1985 and moved to North Carolina. Its newest wave immigrants coming since the 1950s from Puerto Rico suffered the most, not only from racism when competing economically with established immigrant groups (others had come previously from Canada, Poland, Greece, and Ireland) (Beardsley 1993), but also economically when by the mid-1980s many were left without jobs and forced to rely on public assistance (Nunez 2014). This loss in manufacturing and other jobs (a large chicken plant in the area also closed in the late 1970s) has caused Windham County to have one of the highest unemployment rates in the state, as well as one of the state's highest percentages of persons living in poverty, lowest median household income, and lowest per capita income (US Census Bureau 2019), as well as having, as mentioned, its poorest town. Hence, the need for poverty reduction initiatives like CLiCK was great (Godfrey and Freake 2016), even if at the time, it was still only 'a good idea'.

Looking back now, had I not written and received the aforementioned grant, it is possible that CLiCK, as well as the new commitment to poverty reduction through culinary training, food entrepreneurship for those with little to no start-up capital, and accessible/affordable health and nutrition education, might not have emerged, and thereby the Co-op's kitchen project may have remained quite contentedly within the description of the AFM or merely continued as an ongoing verbal plan. Additionally, whether or not Tina and I would have stayed involved is of course open to speculation, for as mentioned, we and some others were already experiencing conflicts within the Co-op (Tina by this time had left the board) around the experiences of some workers and other efforts we were undertaking to make the Co-op more socially inclusive, such as pushing to translate the website into Spanish, proposing the waiving of membership dues based on income, and advocating for a process for workers to address grievances (the Co-op is non-union, as perhaps the assumption is the larger progressive commitments deem one unnecessary). Regardless, we accepted the grant, and by early 2010, CLiCK began to emerge as its own new entity, with new interested people and with $10,000 at its disposal to figure out our next steps.

The new goals were to determine what CLiCK could/should look like to eventually become a non-profit committed in a unique way to the collective tenets of the FJM, even as the non-profit model brings, both in general and to CLiCK specifically, contradictions and challenges to a social justice mission (INCITE! 2007). Additionally, we had no idea how to achieve the goals set forth by the grant, nor any experience doing so, but as the overseer of the grant fortuitously wrote in her email congratulating us, "Ask and you shall receive" (personal communication, July, 2010). This mantra became part of our emerging 'magical' (Abrams

1996) 'abundance mindset' and leadership model (brown 2017; Eisenstein 2011; Freebairn-Smith 2009), one which has continued to be supported and affirmed by events, as in blessings, despite oppositional energies and attitudes. In fact, to end where we began, I have come to understand that the past is, as Barad asserts, never finished, the present and future are not fixed, and the 'magic of beginnings' continues if we embrace the realization that "...change is constant..." (brown 2017, p. 41). This, however, does not mean that change, especially change to promote justice even on the micro scale, is easy or quick. In fact, as will be explored, it is quite the opposite.

> *CLiCK is unique, though. CLiCK offered services with monetary and social value. CLiCK's focused marketing and advertising had a far greater reach than I ever could've done myself. CLiCK helped reduce my costs by offering volunteers so I didn't spend as much on hiring temps. Plus, the support staff CLiCK provided aligned with my company's Corporate Social Responsibility (CSR).*
>
> *CLiCK not only promoted social food justice in the community, CLiCK supported my efforts.*
>
> Middle-class white female, former kitchen member

Notes

1 https://www.azquotes.com/quote/550762
2 See https://native-land.ca/ also see https://accessgenealogy.com/connecticut/connecticut-indian-tribes.htm
3 https://nvdatabase.swarthmore.edu/content/rutgers-university-students-win-divestment-apartheid-south-africa-1985
4 https://dailycampus.squarespace.com/stories/2018/10/5/university-of-connecticut-student-unveiling-history-of-spring-valley-farm
5 https://www.willimanticfood.coop/
6 https://www.nls.uk/learning-zone/politics-and-society/labour-history/fenwick-weavers
7 https://www.willimanticfood.coop/
8 https://ncbaclusa.coop/resources/7-cooperative-principles/
9 See https://www.city-data.com/city/Mansfield-Connecticut.html versus http://www.city-data.com/city/Willimantic-Connecticut.html
10 https://www.congress.gov/bill/111th-congress/house-bill/1/text
11 https://www.covenantsoupkitchen.org/
12 https://viacampesina.org/en/
13 https://www.nps.gov/nr/feature/places/14000434.htm

References

Abrams, D. 1996. *The spell of the sensuous: Perception and language in a more-than-human world.* New York: Vintage Books.
Agyeman, J. 2003. '"Under-participation" and ethnocentrism in environmental education research: Developing "culturally sensitive research approaches', *Canadian Journal of Environmental Education* 8(1): 80–94.

Alkon, A. H. 2012. *Black, white, and green: Farmers markets, race, and the green economy.* Athens: University of Georgia Press.

Alkon, A. H. and Agyeman, J., eds. 2011. *Cultivating food justice: Race, class, and sustainability.* Cambridge, MA: MIT Press.

Barad, K. 2007. *Meeting the universe halfway: Quantum physics and the entanglement of matter and meaning.* Charlotte, NC: Duke University Press.

Beardsley, T. 1993. *Willimantic community and industry: The rise and decline of a Connecticut textile city.* Willimantic, CT: Windham Textile and History Museum.

Bell, M. 2004. *Farming for us all: Practical agriculture and the cultivations of sustainability.* University Park: Pennsylvania State University Press.

Berger, P. and Luckman, T. 1967. *The social construction of reality: A treatise in the sociology of knowledge.* New York: Anchor Books.

Berry, W. 1977. *The unsettling of America: Culture and Agriculture.* San Francisco, CA: Sierra Club Books.

Bradley, K. and Herrera, H. 2016. 'Decolonizing food justice: Naming, resisting and researching colonizing forces in the movement', *Antipode* 48(1): 97–114.

brown, a.m. 2017. *Emergent strategy: Shaping change, changing worlds.* Chico, CA: AK Press.

Butler, J. 1999. *Gender trouble: Feminism and the subversion of identity.* New York: Routledge.

Capra, R. 1997. *The web of life: A new scientific understanding of living systems.* New York: Anchor Books.

Chait, J. 2019. Largest organic retailers in North America. [online] *The Balance: Small Business.* Available at: https://www.thebalancesmb.com/organic-retailers-in-north-america-2011-2538129 [Accessed September 24, 2020]

Christensen, A. D. and Jensen, S. Q. L. 2012. 'Doing intersectional analysis: Methodological implications for qualitative research', *NORA—Nordic Journal of Feminist and Gender Research* 20(2): 109–125.

Currie, G. et al. 2012. 'Institutional work to maintain professional power: Recreating the model of medical professionalism', *Organization Studies* 33(1): 937–962.

DiAngelo, R. 2011. 'White fragility', *International Journal of Critical Pedagogy* 3(3): 54–70.

DiAngelo, R. 2018. *White fragility: Why it's so hard for white people to talk about racism.* New York: Beacon Press.

Eisenstein, C. 2011. *Sacred economics: Money, gift, and society in the age of transition.* Berkeley, CA: North Atlantic Books.

Elliot, T. S. 1917. *Prufrock and other observations.* London: The Egoist.

Elsheikh, E. 2016. 'Dismantling racism in the food system: Race and corporate power in the US Food System: Examining the Farm Bill' [online] Food First. Available at: https://foodfirst.org/wp-content/uploads/2016/06/DRnumber2_VF.pdf [Accessed October 10, 2020]. 2 (summer).

Fairbairn, B. 1994. *The meaning of Rochdale: The Rochdale Pioneers and the co-operative principles* [online]. Center for the Study of Co-Operatives, University of Saskatchewan. Available at: https://ageconsearch.umn.edu/record/31778/?ln=en [Accessed September 24, 2020]

Fannon, F. 2008. *Black skin, white masks.* New York: Grove Press.

Finley, R. 2013. 'A guerrilla gardener in South Central L.A'. Ted Talk. Available at: https://www.ted.com/talks/ron_finley_a_guerrilla_gardener_in_south_central_la. [Accessed 24 September 2020].

Freebairn-Smith, L. 2009. 'Abundance and scarcity: Mental models in leaders'. PhD Thesis. Saybrook University.

Godfrey, P. and Freake, H. 2016. 'Feeding community: A case study of a shared-use commercial kitchen in eastern Connecticut'. In Bosso, C., ed. *Feeding cities: Improving local food access, sustainability, and resilience.* London: Routledge, pp. 113–128.

Godfrey, P. and Torres, D. 2020. 'Situational strangers: Recipes for immigrant lives: Crossing, cooking, cultivating and culture at a shared-use commercial kitchen'. In Agyeman, J. and Giacalone, S., eds. *The immigrant food nexus: Borders, labor and identity in North America.* Boston, MA: MIT Press, pp. 281–298.

Guthman, J. 2007. '"If only they knew": Colorblindness and universalism in California alternative food institutions'. [online]. *The Professional Geographer.* Available at: https://www.tandfonline.com/doi/full/10.1080/00330120802013679 [Accessed October 10, 2020]

Holt-Gimenez, E. and Harper, B. 2016. Dismantling racism in the food system: Food—Systems—Racism: From Mistreatment to Transformation.[online] Food First. Available at: https://foodfirst.org/wp-content/uploads/2016/03/DR1Final.pdf [Accessed October 10, 2020]. 1 (winter-spring).

Holt-Gimenez, E. (2017) *The foodie's guide to capitalism: Understanding the political economy of what we eat.* New York: Monthly Review Press.

Horrigan, L., Lawrence, R. S. and Walker, P. 2002. 'How sustainable agriculture can address the environmental and human health harms of industrial agriculture', *Environmental Health Perspectives* 110(5): 445–456.

INCITE! Women of Color against Violence. 2007. *The revolution will not be funded: Beyond the non-profit industrial complex.* Cambridge, MA: South End Press.

King, T. 2008. *The truth about stories: A native narrative.* Minneapolis: University of Minnesota Press.

Konnikova, M. 2015. 'The real lessons of the Stanford prison experiment', *The New Yorker,* June 12, 2015. [online] Available at: https://www.newyorker.com/science/maria-konnikova/the-real-lesson-of-the-stanford-prison-experiment [Accessed September 25, 2020]

Lawrence, T., Suddaby, R. and Leca, B. 2009. *Institutional work: Actors and agency in institutional studies of organizations.* Cambridge: Cambridge University Press.

Lappé, F. M. 1971. *Diet for a small planet.* New York: Ballantine Books.

Madison, D. S. 2012. *Critical ethnography: Method, ethics and performance.* London: Sage Books.

Magdoff, F., Foster, J. B. and Buttel, F., eds. 2000. *Hungry for profit: The agri-business threat to farmers, food and the environment.* New York: Monthly Review Press.

Massey, D. 2005. *For space: Matter and meaning.* Charlotte, NC: Duke University Press.

Muzika, K. C., Hudyma, A., Garriott, P. O., Santiago, J. and Morse, J. 2019. 'Social class fragility and college students' career decision-making at a private university', *Journal of Career Development* 46(2): 112–129.

Nunez, E. 2014. *Hanging out and hanging on: From the projects to the campus.* London: Rowman & Littlefield.

Patel, R. 2008. *Stuffed and starved: The hidden battle for the world food system.* Brooklyn, NY: Melville House Publishing.

Pellow, D. 2016. 'Toward a critical environmental justice studies: Black Lives Matter as an environmental justice challenge', *Du Bois Review,* 13(2): 1–16.

Pollan, M. 2006. *The omnivore's dilemma: A natural history of four meals.* London: Penguin Books.

Pollan, M. 2013. *Cooked: A natural history of transformation.* London: Penguin Books.

Rabinowitz, A. and Martins, J. 2012. 'Community food security in Connecticut: An evaluation and ranking of 169 towns', *Outreach Reports 154264*, Storrs: University of Connecticut, Charles J. Zwick Center for Food and Resource Policy.

Sandoval, C. 2000. *Methodology of the oppressed*. Minneapolis: University of Minnesota Press.

Shiva, V. 2000. *Stolen harvest: The hijacking of the global food supply*. Boston, MA: South End Press.

Slocum, R. 2007. 'Whiteness, space and alternative food practice' [online] *Geoforum* 38: 520–533. Available at: https://sites.middlebury.edu/gsfswhitepeople/files/2016/09/slocum.pdf [Accessed on September 30, 2020]

Swedberg, R. 2018, 'How to use Max Weber's ideal type in sociological analysis', *Journal of Classical Sociology* 18(3): 181–196.

Tatum, B. D. 2000. 'The complexity of identity: "Who am I?"' In Adams, M., et al., eds. *Readings for diversity and social justice: An anthology on racism, sexism, anti-semitism, heterosexism, classism and ableism*. New York: Routledge, pp. 9–14.

Terkla, D. G. and Pagano, M. F. 1993. 'Understanding institutional image', *Research in Higher Education* 34(1): 11–22.

US Census Bureau. 2019. *2012–2016 American community survey: 5-year estimates* [online]. Available at: https://www.census.gov/programs-surveys/acs/technical-documentation/table-and-geography-changes/2016/5-year.html.[Accessed September 25, 2020].

White, M. M. 2018. *Freedom farmers: Agriculture resistance and the black freedom movement*, Chapel Hill: The University of North Carolina Press.

Zitcer, A. 2017. 'Collective purchase: Food cooperatives and their pursuit of justice'. In Alkon, A. and Guthman, J., eds. *The new food activism: Opposition, cooperation, and collective action*. San Francisco, CA: University of California Press, pp. 181–205.

5   Development of CLiCK from within and without

We are in an imagination battle.

adrienne maree brown

An insider/micro perspective

One of the challenges in creating CLiCK (Commercially Licensed Co-operative Kitchen) or any social justice project, initiative, or organization is that we must, as brown observes, engage over the long term "…in an imagination battle" (2017, p. 18), while recognizing that new models with replicable results are hard to find. In fact, as Neva Hassanein (2003) astutely argues, "…calls for fundamental change and complete transformation of the agro-food system are rarely—if ever—accompanied by specific suggestions on how to achieve such a total makeover" (p. 84). Additionally, given the corporate food system's global dominance from production to consumption (Patel 2008), any efforts of imagining, nurturing, and securing alternative successful food system models (local, regional, or global) are consequently under extreme economic and ideological duress, making their very existence an ongoing 'battle'. Consequently, many alternative food ventures totally fail not long after beginning, and/or they fail in making any significant changes by only addressing consumer *choices* as with the Alternative Food Movement (AFM), as opposed to addressing the very structures of the system, as with the Food Justice Movement (FJM). As such, many predicted that CLiCK would totally fail – including our first major funders – and it is still possible we will. Yet, as wisely understood by Buckminster Fuller who stated, "You never change things by fighting the existing reality. To change something, build a new model that makes the existing model obsolete" (Sieden 2011, p. 358.), it is only by changing our 'realities', hence our 'models', that authentic change can actually occur. Thus, for me a key part of the FJM, which I am taking to include the ideas and ideals of just sustainabilities (JS), is exploring new ideas and new ways of relating to our food and to others that simultaneously challenge the structural and social inequalities of the existing society/food system. Seminal among such new ideas and practices is the concept of food democracy, understood here as an informal 'political' organizational method (model) for achieving the social goals of justice

(Miller 1978), in this case, specifically those of JS. As the global food activist and writer Vandana Shiva (2016) states, "…food democracy ….is the new agenda for ecological sustainability and social justice" (p. 18), whereby peoples of the world have collective cultural control over their food, including over their rights to save, select, and plant seeds without corporate interference or profiteering. Additionally, for Hassanein (2003), "The concept of food democracy rests on the belief that every citizen has a contribution to make to the solution of our common problems" (pp. 84–85) – if, I would add, they are given the opportunity to do so in ways conducive to their needs, skills, and abilities. Moreover, such efforts at engaging food democracy must, as Hassanein (2008) also identifies, involve "…collective action by and among organizations" (p. 290), as well as by individuals who can play roles "…in governing and shaping their relationships to food and the food system", thereby "…gaining knowledge, sharing ideas, developing a sense of efficacy, and contributing toward the community good" (p. 295).

As such, in the early days of CLiCK, in addition to procedural challenges, we also needed to find the people who would be these 'citizens' interested in contributing 'to the solution of our common problems'. Availability of funding made it easier to garner interest, as we had seed money to support research into our community's needs and to explore what others in the region were doing. As mentioned, although Tina and I had originally been motivated by the tenets of the AFM and then by those of the FJM, coinciding with my social justice commitments as a sociologist, we had no personal *need* for CLiCK – we didn't want to start a food business, we weren't farmers, and we weren't food insecure or in need of nutritional education. However, we did want to continue serving 'our community', while also extending the reach of that 'community' to include more racial, social class, and cultural diversity, engaging what Kelly Moore and Marilyn Swisher (2015) call "… an empowerment model" (p. 117). Hence, we needed to move beyond the small farmer needs identified by the Co-op to find out who else might need CLiCK, how might they need it, why might they need it, and under what conditions might they turn that need into action. To do so, we had to engage in research, investigating the needs in our local community but also learning about what others were doing elsewhere, how they were doing it, and what challenges lay ahead. With the help of a graduate social work intern, we created surveys in the spring of 2010 for different local actors (farmers, potential entrepreneurs, potential students, etc.) and made plans to visit examples of similar projects in New England over the summer of 2010. We also looked to the Community Food Security Coalition (CFSC) for guidance, as at the time (it folded in 2012, as in the earlier point about failure), it was "…the dominant private, nonprofit organization in the field of community food security" (Morales 2011, p. 152). The CFSC held a conference in the fall of 2010 in New Orleans and so we used grant funds (as approved by the project administrator) to attend, along with two other committed individuals – both white females (one a nutritionist, who become a long-term board member, and the other a single-mother, who was considering starting a food business but never did).

These early efforts proved to be very instructive. From the surveys, we learned that the potential CLiCK project had local interest and support, not just from the original farmers but also from potential culinary entrepreneurs, nutrition educators (including the one already mentioned), after-school programs, etc. Based on site visits to a small farmer's commercial kitchen in Amherst, MA, and a shared kitchen in Portland, ME, called Local Roots, which has since closed (another failure), we learned that such a project would be extremely costly, politically challenging, and difficult to achieve. The CFSC conference further emphasized that such a project would have to first and foremost be *community* based and supported, as well as culturally inclusive (including race and class). Additionally, despite the benefits of learning from others, we had to accept, as Morales (2011) argues, that "…each social, environmental, ecological, economic, and political matrix of race, [class] and food is unique" (p. 169). In other words, as quoted already – 'we would make the road by walking' – or perhaps more accurately in our case, by stumbling! To avoid *excessive* stumbling, the logical next step seemed to be to take the grant deliverable – our completed business plan that pulled together all our findings – and formalize it by making CLiCK a 501c3 non-profit, with the help of those interested in serving on the proposed new board.

Having finished with our first grant, we were somewhat aware (soon to become more so) of the non-profit trap in terms of the restrictions grant funds can bring, especially in relation to social justice initiatives (INCITE! 2007). Yet, we were also aware that without 501c3 legal status, future grants would be hard to access, so we sought to find a compromise by maintaining CLiCK's cooperative roots, even though non-profits cannot distribute profits to members as is part of the cooperative model. We sought legal assistance from a community member on the faculty at University of Connecticut's (UCONN's) Law School to help us draft our by-laws and articles of incorporation that would enable us to create different categories of 'membership', hoping to ensure that those involved with CLiCK would have a voice and thereby a sense of ownership. Additionally, we sought to base CLiCK on the seven cooperative values that we felt would help us stay committed to our originally funded objective – to address local poverty through innovation in a manner based on justice, equity, and empowerment, rather than just charity.

By creating the possibilities for individual empowerment through membership that would include access to the commercial kitchens, as well as potential access to job training, educational offerings, and other social services, we ended up combining aspects of the two general classifications of non-profits – mutual and publicly orientated (Quarter et al. 2001). The basic difference is that,

> In mutual nonprofits, the members are the recipients of the service…. revenues come from members either through an annual fee or a fee for service, and the members have the right to participate in the affairs of the organization through electing representatives to the board of directors.
>
> (p. 352)

Generally speaking, "Mutual nonprofits, … are inwardly oriented, whereas other forms of nonprofits are outwardly oriented, either to the public at large or to a specific public (for example, people living in poverty or people with a particular affliction)" (p. 352). In attempting to once again bridge a divide (as in between the AFM and the FJM) and become both types of non-profit, while upholding the co-operative values, CLiCK faced a conceptual challenge when writing our by-laws/ articles of incorporation and a categorical challenge with future funders, as will be later explored. It is worth noting, however, that by bringing together different organizational approaches, we sought to balance creating a sense of solidarity through membership, while still being outwardly oriented in our commitment to the larger community, as well as still sticking to the seven cooperative principles. We thereby sought to avoid the "…primordial or ascriptive principles of membership" that can "…be racist … anti-democratic … and elitist …." (Seligman 1998, p. 80), as evident in many aspects of the AFM. Moreover, in combining approaches, we sought to increase the chances of CLiCK being able to address economic and social inequality through embodying what McLaren and Agyeman (2015) have termed the …"solidarity economy…", characterized by "…spaces, services, and goods…" being created "…*with* rather than *for* users, their families and neighbors" [italics in original] (p. 11). In other words, we sought to collectively empower users and members of CLiCK and our larger community, rather than merely *serve* them as in the more typical charity non-profit model.

By this point, we now had on our emerging board (of which I was the president) several individuals from diverse social classes and employments (a UCONN cook, a supermarket baker, a nutritionist, and a culinary educator), all of whom were all interested in using the future kitchens (commercial and/or teaching) themselves once they existed. The criteria for CLiCK's board was to have officers who could commit a significant amount of time and energy to support our cooperative values and commitment to social justice. In addition, and as per our by-laws, our board would have a kitchen member representative who would represent their issues and help to organize an annual membership meeting. We never practiced the board dictum of 'give, get or get off', but later on, we did ask that board members become members of CLiCK and thereby give membership dues, although we never enforced it. In fact, even today, the board remains rather unstructured in relation to term-limits and protocol partly because we sought to create an inclusive, flexible, and individualized environment. This style of leadership and operations was challenging for many, as will be discussed more later, but the principle was to encourage shared power and responsibility. Although most of these individuals were white and female, our vice president was a Latina from Puerto Rico who worked for UCONN Extension in a multitude of roles, including as a nutrition educator in Windham County and also worked with the national organization Cooking Matters.[1] Through her work, including highlighting the need to translate our promotional materials into Spanish (as well as the need to do it well, while pointing out that depending on the country of origin of whomever is doing the translation, a different dialect will be evident and influential as to whom it speaks to), she brought other interested individuals to the project, both from the Latinx community (including those from Puerto Rico, Mexico, and Guatemala who generally do not see themselves as a 'single community', as will be later explored) and from the dominant white one. As such, we were beginning to see increasing racial, cultural, occupational, political, and social class diversity that brought with it a sense of achievement, in that it felt as if we were already making progress in our 'imagination battle' for inclusive spaces. Additionally, as CLiCK's board members and supporters, our emerging diversities were experienced through our shared mission and collective visioning, allowing us to recognize that our means to achieving our goals were inseparable from our ends. This was an insight I had garnered from the Occupy Movement and their use of General Assemblies, where consensus was used both as their ends and their means (Milstein 2010), thereby affirming Thich Nhat Hanh's observation that "...there is no way to happiness, happiness is the way".[2] This new diversity would continue to enrich and challenge CLiCK's emerging identity in surprisingly complex ways that, although unexpected at the time, upon critical reflection did follow predictable social patterns. For as much as CLiCK sought to put aspects of the cooperative values and consequently aspects of JS into practice, it is important to note that our efforts were not at all conflict-free or fully successful, as will be further explored.

> *I think sustainability is one of the top priorities for CLiCK because social and food justice would be nothing without it....I would love to see growth in the teaching section of CLiCK where classes were connecting their lessons to climate change in daily, weekly, or monthly discussions. It is very important to make this clear, in multiple languages, so that CLiCK can become a reliable resource of extensive knowledge to the community. The population of our city has a strong Latino presence, one that is in danger of having very bad health problems due to food deserts/food apartheid.*
> Middle-class Latina, former CLiCK intern

An outsider/macro perspective

By spring of 2011, CLiCK was an official non-profit incorporated under section 501c3 of the Internal Revenue Code, with cooperative-based by-laws that highlighted the role of 'members'. There would be three classes of members – individuals, other non-profits, and institutions – which would have a say in aspects of CLiCK's operations, be represented on our board, and receive some form of benefits. As a new non-profit, CLiCK would be joining 1.5 million others in an ever-increasing sector of the economy (McKeever 2018). As a result, CLiCK was now part of the 'non-profit industrial complex' (INCITE! 2007), as well as what Paul Hawkins (2008) referred to as the "...largest social movement in all human history" (p. 4). Such a movement, Hawkins imagines, is akin to the Earth's immune system rising up from the grassroots to defend itself from ecological destruction. This is a powerful image, but as Troy Wiley (2018) notes in his blog for *World Summit*, a body's immune system works in unison, whereas, despite Hawkins' desire to see non-profits around the world as all working *together*, in many cases, they are competing for "…the same charity dollars that are increasingly in scarce supply as socio-economic conditions get worse" (para. 5). Additionally, as Wiley notes, many such publicly orientated organizations are consequently forced to merely address symptoms of injustice, rather than the root causes of structural inequalities created by the complex intersections of global patriarchy, colonialism, white supremacy, and capitalism (Holt-Gimenez 2017; Patel 2008). In contrast, our vision for CLiCK as a shared-use kitchen based on cooperative values, focused on enabling those of limited economic means to open food businesses, including food trucks and carts (Godfrey 2017), was that it would engage the tenets of community economic development (Agyeman 2013). By using a model that is supportive of individual businesses, while remaining rooted in the 'solidarity economy' (McLaren and Agyeman 2015), CLiCK aimed to address root causes of inequality, in this case of *entrepreneurial* inequality, as in unequal and limited access to start-up capital (some of our grants, as well as some of our fundraising, provided start-up capital to eligible individuals) and to the means of production, in this case represented by its fully equipped commercial kitchen.

Yet, in order to achieve these lofty goals, CLiCK had to rent, buy, build, or otherwise access an *actual* commercially licensed kitchen, as well as teaching space, that would be big enough and well equipped enough to serve the needs of many different, yet unpredictable, future culinary entrepreneurs, including farmers. For a new non-profit with no funds, this was of course a seemingly impossible task, and so at this time, all we could do was imagine and dream, while making those dreams publicly known. As a sociologist, I know that humans construct their realities by repeatedly manifesting them into existence, recalling the ideas of the sociologists Berger and Luckman who state that, "The most important vehicle of reality-maintenance is conversation…" that both "…ongoingly maintains reality, …[and] ongoingly modifies it" (pp. 172–173). Additionally, I knew that my role as a sociologist carried a "…privileged status…" (p. 174) that would give CLiCK's narrative some plausibility – helping community members to believe we would make it happen – despite the obvious challenges. In fact, time and time again, when we would tell people about the idea of CLiCK, they would respond with enthusiasm; the idea alone seemed to consistently inspire in people a sense of what environmental/spiritual author Charles Eisenstein has termed "the more beautiful world our hearts know is possible" in his book with this title (2013). In other words, I learned that we needed to 'reality-construct' by repeatedly 'conversing' CLiCK into existence, doing so in such a manner that would create an inclusive, inspiring, grand narrative capable of reaching people's desires for new possibilities and ultimately for hope. During the time that CLiCK did not physically exist, we still needed people to think and act as if it *did*, while at the same time, we also needed to believe that if we focused our intention (McTaggart 2008) and our efforts enough, the 'universe would provide' and thereby we would collectively shape, as proposed in the Introduction, our JS imaginary. The role of my sociological views on reality construction, as well as Tina's and my spiritual beliefs in 'abundance', not to mention our agreement with M. L. King Jr.'s vision that "…the arc of the moral universe is long, but it bends towards justice" (1968) helped us embrace the unknown, engage in "…self-deception" and thereby remain optimistic, hence "positive" (Sheridan et al. 2015), despite the very realistic overwhelming financial and social odds. Looking back, what Tina and I were doing (I can't speak for our other board members) can be also be understood as "engaged spirituality" (Stanczak 2006), in that our physical labor (emails, meetings, promotion events, etc.) was accompanied by our intense intention setting and, in my case, prayer. Such prayer did not solicit intervention from the usual suspects but rather from the 'mystically moral universe', which is made apparent to me by the many splendid manifestations of life itself, as well as in relation to CLiCK, not to mention my myriad day-to-day inexplicable experiences.

By 2012, our board had eight members and we were meeting monthly at the Covenant Soup Kitchen[3] (already mentioned in relation to the original grant), a place that through donations and volunteer labor provides 12 meals a week to hundreds of low-income people in need. To do so, they have a small commercial kitchen, and we considered it as an option, along with other church and senior center kitchens, but the issues of size and lack of commercial equipment were problematic. We also spent months in conversation with the social service agency from whom we had received our original grant to see if some of the extra space they had might work, while linking CLiCK with possible plans for a new senior center (finally under construction) and housing (has yet to happen). There were other ideas and site visits, but none of them had the space and the potential for fulfilling our mission, nor did any of them have a way to both serve culinary entrepreneurs and offer health and nutrition education and outreach. As a result, we realized CLiCK would need its own building if it were to successfully carry out its mission. In the fall of 2011, to gain new ideas and insights, we organized a visit to tour the Franklin County Community Development Corp (CDC) Food Processing Center in Greenfield, MA (see Appendix), which is a shared-use commercial kitchen that rents space, does its own processing for local farmers, and sells produce to a number of schools/universities in their area.

This tour was the first time we were able to see in person an example of what CLiCK could look like and fully realize the extent of the necessary financial and logistical undertaking. At the same time, seeing kombucha being made, local apples being processed, and hot sauce getting ready to be bottled, ignited a sense of excitement and possibility in us. In referring in the last chapter to the possible reasons why, based on my life story, I may have taken up this project, one that I did not mention was my childhood fascination with Roald Dahl's children's book *Charlie and the Chocolate Factory* (1964). Visiting the Franklin County CDC gave me a feeling of excitement held over from Dahl's 'imaginary chocolate factory' – as this was a place where things were *made*, evoking a sense of intense creativity and even alchemy. In college, I had been fascinated by Carl Jung's work on alchemy and the collective unconscious (Adler and Hull 1968), which at this time fed into my sense that thoughts and their manifestations are somehow cosmically connected beyond the realm of our understanding. Hence, it seemed to me that the thoughts in my mind, if correctly 'cooked', could be transformed into new and edible (visible) matter, becoming much more than the sum of their original 'ingredient' parts.

And so, the months turned to years, and we continued to search for a building, gather ideas, and make public our vision through small-scale fundraising events and community outreach, all the while believing that eventually, somehow, it would happen. Then, in 2013, I received an email from a community member letting me know that the Knights of Columbus building in Windham was up for sale and that there was a commercial kitchen in it, as well as 5,600 sq feet of space and 2.6 acres of land. At the time, Tina and I had recently had our roof fixed by a white male neighbor who owned his own construction company and so when setting up an appointment to view the building we asked him to come along to offer us his professional opinion.

My first impressions of the building were mixed. It had been built in the 1970s and was energetically stale, as in the overwhelming smells of beer and cigarettes still emanating from the 'old men's bar' equipped with pool tables, dart boards, televisions, and a peeling pizza restaurant type wall image of Italy. The commercial kitchen was dirty and decrepit, complete with a few 'pin-up girls' and the banquet room – along with the giant electronic bingo board counter on the wall and the cheap chandeliers – was depressing. Yet, the space clearly had potential. I could right away see through these thick shadows of the building's past incarnation; so could Tina and, more surprisingly, so could our neighbor. In fact, no sooner had the real estate agent finished giving us the tour than our neighbor called me aside to talk. We stepped outside, and as if speaking from the mouth of destiny, he said, "CLiCK should buy this building and I will provide the funds in the form of a mortgage". Now, as mentioned, I had been asking the universe for such a moment, but I never expected it to come in the manner it did, nor from the person it did, nor did I know what to say when it did. I don't even remember what I said, except maybe, "Wow, thank you" and then "Are you sure?", to which he replied something along the lines of "I have never done anything for my community and your idea and your commitment to it has inspired me to want to help". And he did.

The building was purchased by CLiCK in fall of 2014 (see Figure 5.1) and, as a result of the funds put up for the mortgage, CLiCK was now eligible for a

Figure 5.1 Building and Sign.

matching grant of $100,000 from the Community Economic Development Fund (CEDF),[4] a non-profit that receives funds from the CT Department of Economic and Community Development (DECD)[5] for small businesses and, in our unique case non-profits, since CLiCK's plan was to be a small culinary business incubator. We had learned about these funds because we had already been working with CEDF's board President (a white woman), who, as it turned out, was good friends with the organization's CEO (another white woman), who affirmed our mission by declaring CLiCK the "Best idea I've seen in years". This statement not only helped instill confidence in our mortgage provider, but was also notable for marking how radically her sentiment would later change. However, the social connection between the board president and the CEO was my first insight into how state and local funds are distributed through social and political channels that follow an internal logic all their own. In fact, despite feeling extremely grateful and spiritually energized – having now not only bought a building due to the 'kindness of a stranger' but also having received a $100,000 matching grant – it was while sitting in the boardroom of CEDF in Meriden (a room decorated with all their bank and corporate supporters) signing a pile of legal documents that I began to feel newly anxious. At first, I thought it was just the stress of increasing financial responsibility as CLiCK's board president, a common sentiment among non-profit leaders (Kanter and Sherman 2016; McDonald 2013). However, later, it came to light that the seemingly entirely beneficial arrangement was not quite as it had appeared, and we were, as it turned out, also being subtly misled and used by the other non-profit for its own agenda. Whether it was this yet to be disclosed subtext that somehow added to my anxiety, or whether it was the additional physical and emotional labor now required to carry through on our financial commitments, I cannot determine even now. Regardless, for the next few years, I would develop what Tina and I privately came to refer to as my 'CLiCK-ease' – the stress and 'dis-ease' that came from all the responsibility and work such an undertaking involved. In fact, had we not had each other, I don't think I could have withstood the stress I began experiencing, as if the success of CLiCK had now become dependent on my personal ability to make it happen. Of course, this was *not* the case, as we were in it together along with many others who were now involved, and yet this impression persisted, at least if our vision based on social justice was to remain and somehow become actualized. I state this because in the everyday overwhelming logistics of running a non-profit, attention to the more idealistic and intangible goals such as of achieving JS can easily get sidelined, if not continually and intentionally focused on, as will be explored next.

Sustainability is about regenerative practices for the Earth/soil, people/neighbors, and local businesses and local business opportunities. CLiCK supports and nurtures all!

It is a very inspiring example of what can happen with good ideas when communities come together with a common vision.

Middle-class white female, community partner through local university

Notes

1 https://cookingmatters.org/
2 https://www.wonderfulquote.com/a/thich-nhat-hanh-quotes
3 https://www.covenantsoupkitchen.org/
4 https://www.cedf.com/
5 https://portal.ct.gov/DECD

References

Adler, G. and Hull, R. F. C., eds and trans., 1968. *Collected works of C.G. Jung, volume 12: Psychology and alchemy*. Princeton, NJ: Princeton University Press.

Agyeman, J. 2013 'From loncheras to lobsta love: Food trucks, cultural identity and social justice' [online]. *julianagyeman.com*. Available at: http://julianagyeman.com/2013/06/from-loncheras-to-lobsta-love-food-trucks-cultural-identity-and-social-justice/. [Accessed September 25, 2020]

brown, a. m. 2017. Emergent strategy: Shaping change, changing worlds. Chico, CA: AK Press.

Dahl, R. 1964. *Charlie and the chocolate factory*. New York: Alfred A. Knopf.

Eisenstein, C. 2013. *The more beautiful world our hearts know is possible*. Berkeley, CA: North Atlantic Books.

Godfrey, P. 2017. 'Reflexive food-truck justice: A case study in CLiCK, Inc, a non-profit shared- use commercial kitchen'. In Agyeman, J. Matthews, C. and Sobel, H., eds. *Food trucks, cultural identity, and social justice: From Loncheras to Lobsta Love*. Cambridge, MA: MIT Press, pp. 149–167.

Hassanein, N. 2003. 'Practicing food democracy: A pragmatic politics of transformation', *Journal of Rural Studies* 19: 77–86.

Hassanein, N. 2008. 'Locating food democracy: Theoretical and practical ingredients', *Journal of Hunger and Environmental Nutrition* 3(2–3): 286–308.

Hawkins, P. 2008. *Blessed unrest: How the largest social movement in history is restoring grace, justice, and beauty to the world*. London: Penguin Books.

Holt-Gimenez, E. 2017. *The foodies guide to capitalism: Understanding the political economy of what we eat*. New York: Monthly Review Press.

INCITE! Women of Color against Violence. 2007. *The revolution will not be funded: Beyond the non-profit industrial complex*. Cambridge, MA: South End Press.

Kanter, B. and Sherman, A. 2016. *The happy healthy nonprofit: Strategies for impact without burnout*. New York: Wiley.

King, M. L. 1968. 'Remaining awake through a great revolution. Speech given at the national Cathedral, March 31. [online]. Available at: https://www.si.edu/spotlight/mlk?page=4&iframe=true [Accessed September 30, 2020]

McDonald, G. 2013. 'Stressed out?' [online] *governing good.ca*. Available at: http://www.governinggood.ca/stressed-out-2/ [Accessed September 25, 2020]

McKeever, B. 2018. 'The nonprofit sector in brief 2018: Public charities, giving, and volunteering' [online]. *National Center for Charitable Statistics, Urban Institute*. Available at: https://nccs.urban.org/publication/nonprofit-sector-brief-2018#the-nonprofit-sector-in-brief-2018-public-charites-giving-and-volunteering. [Accessed September 25, 2020]

McLaren, D. and Agyeman, J. 2015. *Sharing cities: A case for truly smart and sustainable cities*. Cambridge, MA: MIT Press.

McTaggart, L. 2008. *The intention experiment: Using your thoughts to change your life and the world*. New York: Atria Books.

Miller, D. 1978. 'Democracy and social justice', *British Journal of Political Science* 8(1): 1–19.

Milstein, C. 2010. *Anarchism and its aspirations*. Chico, CA: AK Press.

Moore, K. and Swisher, M. E. 2015. 'The food movement: Growing white privilege, diversity, or Empowerment?' *Journal of Agriculture, Food Systems and Community Development*, 5(4): 115–119.

Morales, A. 2011. Growing food *and* justice: Dismantling racism through sustainable food systems. In Alkon, A. H. and Agyeman, J., eds. *Cultivating food justice: Race, class, and sustainability*. Cambridge, MA: MIT Press, pp. 149–176.

Patel, R. 2008. *Stuffed and starved: The hidden battle for the world food system*. Brooklyn, NY: Melville House Publishing.

Quarter, J., Sousa, J., Richmond, B. J. and Carmichael, I. 2001. 'Comparing member-based organizations within a social economy framework', *Nonprofit and Voluntary Sector Quarterly* 30(2): 351–375.

Seligman, A. 1998. 'Between public and private: Towards a sociology of civil society'. In Hefner, R., ed. *Democratic civility*. New Brunswick, NJ: Transaction, pp. 79–111.

Sheridan, Z., Boman, P., Mergler, A. and Furlong, M. J. 2015. 'Examining well-being, anxiety, and self-deception in university students'. [online] *Cogent Psychology*, 2(1). Available at: https://www.tandfonline.com/doi/pdf/10.1080/23311908.2014.993850 [Accessed September 25, 2020]

Shiva, V. 2016. *Earth democracy: Justice, sustainability, and peace*. Berkley, CA: North Atlantic Books

Sieden, L. S. 2011. A fuller view – Buckminster fuller's vision of hope and abundance for all. Divine Arts Media.

Stanczak, G. C. 2006. *Engaged spirituality: Social change and American religion*. New Brunswick, NJ: Rutgers University Press.

Wiley, T. 2018. 'Critical mass and the blessed unrest that never found its power…Until Now' [online]. *Medium.com*. Available at: https://medium.com/world-summit/critical-mass-and-the-blessed-unrest-that-never-found-its-power-until-now-f6d500fc75e7 [Accessed September 25, 2020]

6 Institutionalization of CLiCK from within and without

The universe is change, our life is what our thoughts make it.

Marcus Aurelius

Institutionalization of CLiCK (Commercially Licensed Co-operative Kitchen) from within

Although I ended the last chapter by discussing my emotional stress from the increased level of responsibility I felt from CLiCK, I want to start this one by highlighting that, regardless, Tina and I were able to maintain a high level of optimism that the universe would support our thoughts, intentions, and efforts to create positive social change. This positive mindset is one we share in all things but now that we had been unexpectedly able to purchase a building, we felt it was less based on our 'self-delusion', as mentioned, and more based on hard evidence that change is possible or rather that it is inevitable. For as the Roman Emperor and philosopher Marcus Aurelius identified, in a sentiment later echoed by Octavia Butler that, "The only lasting truth is change" (1993, p. 3), that the universe itself – or God, depending on which you choose – is *change*. Tina and I prefer the language of 'the universe', but we still like to engage in forms of divinization by pulling tarot cards, using a set based on Celtic pagan teachings (Franklin 2002). When we were considering purchasing the building, we decided to consult the cards, which can be seen as further evidence of our 'self-delusion' or of our 'magical' sensibility or of both. Together, we pulled a card from the Major Arcana, #21, *The World Tree*, which represents the *axis mundi*, linking all three realms – the Underworld (the pagan realm of the Goddess, the earth mother, as opposed to the Christian notion of hell), the Middle Earth (where we live), and the Heavens (home of the gods, as opposed to the Christian notion of heaven). According to our set, *The World Tree* suggests "...the successful completion of some venture, achieving something that you have worked hard for: the successful end of a cycle of events" (Franklin 2002, p. 144). Additionally, on a more spiritual level, it represents "...the Oneness and interconnection of all things, their equal importance and their myriad possibilities" (p. 144). Needless to say, we felt vindicated by this card. As a result, both from the specific chain of events leading to the selection and purchase of CLiCK's building and from this card's timely appearance, we really did feel that everything was interconnected and that anything was possible if we continued to stay focused with our thoughts and our intentions. In short, as Maya Angelou advises, "Ask for what you want and be prepared to get it"[1] and so we were preparing to get that for which we had asked.

Having asked for years for a building and now having finally received one, along with $100,000 to fix it up and create a commercial kitchen, we then had to begin a whole new type of labor. Previously, our labor was more symbolic – doing outreach, sending emails, going to meetings, and creating the impression that CLiCK existed, even though it didn't yet, outside of our 501c3 legal standing and our vivid imaginations. Now, the board had voted to purchase a building, giving us a physical space that could be changed through our physical labor into an actual place – a place called *CLiCK*. In addition, the individual who had loaned CLiCK the funds for the building mortgage asked to join the board so that he might support its future and no doubt safeguard his funds. At times, this situation seemed to present a conflict of interest, especially when CLiCK's financials and future became increasingly precarious, thereby delaying his monthly payments, but at others, it added a much needed business background and a fiscal conservative to our board mix.

One of my great intellectual passions since graduate school has been for the early philosophical works of Karl Marx and the insightful analysis offered by Erich Fromm's book *Marx's Concept of Man* (1974). In fact, I have previously written about applications of Marx's theory of alienation/estranged labor in the current educational system in the USA, arguing that students, just like Marx's workers, are alienated from their learning/labor (Godfrey 2017a). My familiarity with these works has also influenced my approach to labor at CLiCK – the labor to get CLiCK off the ground, the ongoing labor to keep it running, and the labor of the emerging entrepreneurs who choose to start culinary businesses through use of the shared kitchens.

For Fromm (1974), what has been most misunderstood about Marx is that his 'materialism' has been seen as "anti-spiritual"; Fromm refutes this, arguing that,

> Marx's aim was … the spiritual emancipation of man, of his liberation from the chains of economic determinism, of restituting him in his human wholeness, of enabling him to find unity and harmony with his fellow man and with nature.
>
> (p. 3, quoted in Godfrey 2017a, p. 97)

To articulate this concept of 'human wholeness', Marx proposed the concept of *Gattungswesen*, translated as "species-being". My interpretation of species-being, like that of James M. Czank (2012), is that it represents an expression of our "… universal nature…" (p. 318; quoted in Godfrey 2017a, p. 92), our innate *creative and spiritual capacity* to socially construct unlimited cultural realities that include everything from languages, to religions, to rituals, to cooking, to clothing, to buildings, to institutions, and more, all of which constantly pull from, engage with, and transform our natural surroundings to varying degrees. Nick Dyer-Witheford further articulates our species-being as "…the social elaboration and expansion of … life-needs …[entailing]… material capacity, self-consciousness, and collective organization, all feeding into each other" (Dyer-Witheford 2004, p. 476; quoted in Godfrey 2017a, p. 92). As such, Marx's critique of capitalism gains poignancy in that under capitalism the laborers – and consequently, I would add, most of society – are estranged from their creative and spiritual capacities, their own labor power. As a result, most of us have little or no time or adequate skills to make that which we want or need in the manner which we would choose (Godfrey 2017a). As a result, in this hyper-capitalistic society, most of us have become, to paraphrase Marx, alienated from nature, from ourselves and from our own productivity/creativity and ultimately from our species (as cited in Fromm 1974, p. 101).

In fact, in many ways, had it not been for CLiCK and its increasing material manifestation, I would not have been able to, as Marx says "…actively and in a real sense, and sees his [my/our] own reflection in the world which he has [I/we had] constructed" (Marx 1844, para. 32). Once CLiCK purchased its building, I was increasingly able to see our vision grow into an institutional reality, which as mentioned in the last chapter, has become an alchemical process that itself mimics cooking. Furthermore, it is not only my own 'reflection' that has been 'constructed', but also the reflections of those who have served on CLiCK's board. In addition, as a culinary business incubator, CLiCK has made it possible for others to engage their 'species-beings' and make their own reflections grow, while also potentially 'restituting their human wholeness' and even finding 'unity and harmony' with others working there, as will be further explored. But I am getting ahead of myself, for after purchasing the old Knights of Columbus building at 41 Club Road in Windham, CT, there was much to be done before any actual cooking could take place there. Still, all along the way to achieving that long-term goal, I could feel my 'species-being' nonetheless being fed.

The building we purchased had been left full of stuff from over 30 years of use by the Knights of Columbus, and so, it took about six months of work just to clean it out, do minor repairs, and begin to visualize how to turn the old decrepit kitchen into something that could be licensed, rented, and made desirable to future culinary entrepreneurs. Given that we now had $100,000 in grant money from Community Economic Development Fund (CEDF), we could temporarily hire someone to coordinate the logistical steps and expenditures of the proposed kitchen build out. Tina, who had extensive administrative experience in the medical, construction, and entertainment businesses and who had flexibility in her schedule, took on the role, thereby stepping off the board. In addition, Tina and I were very adamant that the property must reflect our commitment to not only social justice but also to a sense of creativity, abundance, and inclusion. To achieve this, we worked with the rest of the board to clean out the inside of the building, and individually and with our personal friends, as well as other volunteers, we also worked on the grounds to plant gardens and a 30-tree orchard with trees donated by a local wholesale landscape company (see Appendix about the current state of these 30 orchard trees), create a sacred labyrinth (a heart-shaped one designed by a friend of ours), secure two bee hives (overseen by another friend of ours who is also our car mechanic), and install a sign based on our logo (designed by a University of Connecticut [UCONN] art student). Given that we had the soil around CLiCK tested and found that it had a high level of lead (the Knights had not been good land stewards), we invited a farmer certified in permaculture (the ethics of which are 'Earth Care, People Care and Fair Shares'[2]) to assist us in the planning of the gardens and orchard. As such, for the gardens we used what is known as hügelkultur, that means using wood and leaves to build up the soil, supplying the plants with ongoing nutrients and trapping in moisture (Miles 2010). This has proven very successful as the gardens and the orchard (also planned based on permaculture methods) have thrived, despite natural challenges, such as three years of gypsy moths who love apple trees. In later years, we added a mural[3] (see Figure 6.1) to one side of the building featuring the food practices of the local Native American tribes (as in hunting and gathering, as well as farming, including the 'three sisters' – corn, beans, and squash), a fire pit, and we expanded and fenced in the raised vegetable garden with the help of a UCONN student who had received an IDEA grant, to increase community engagement with the gardens.

Since no one on the board had any experience in designing, building, or running a commercial kitchen, our first task was to consult those who did. One of the pioneering shared-use commercial kitchens in the North East, The Vermont Food Venture Center (VFVC), was started by the Center for an Agricultural

Figure 6.1 Native American Foods Ways Mural.

Economy (CAE), a 501c3 non-profit founded in 2004 in Hardwick, Vermont (see Appendix). The story of Hardwick and its small-scale local food rebirth is captured by author Ben Hewitt in his book *The Town that Food Saved: How One Community Found Vitality in Local Food* (2009). Hewitt explores the possibility that Hardwick "…just might prove what advocates of a decentralized food system have been saying for years: that a healthy agriculture system can be the basis of communal strength, economic vitality, food security, and general resilience in uncertain times" (p. 2). Of course, this is how it *used* to be for some to a certain extent around the country, although our 'decentralized food system' of the past was never socially *just* given the violent removal of Native Americans, the reliance on slavery, the oppression of women, and other aspects of our society's exploitative past that still continue. Hewitt's proposal that Hardwick could represent a model to help awaken the country from the "…spell of industrial agriculture…" (p. 222) has some merit, but unfortunately lacks an adequate recognition of all the intersecting inequalities, which must also be named and confronted if we are to truly address the *full* dysfunction of our corporate food system. However, for my purposes here, Hardwick's recognition of the role of the VFVC is highly relevant; he describes it as a "…place for small-scale producers to create and distribute value-added goods made with local ingredients, saving them the massive expense and hassle of installing such a facility on their own properties" (p. 3). This shared-use kitchen model was our plan as well, and therefore, we looked to the VFVC for support in developing it, ultimately hiring one of the founders, Brian Norder[4] (a white male), to come and share his expertise.

Brian's visit and expert consultation with us ironically led to an increasingly dire conflict with CEDF, who held our funding, a conflict which then spilled out into our board. The conflict emerged from Brian's proposal to not only completely redo the existing kitchen in our building but also build a second kitchen so that they could be rented by two users at the same time. These proposed kitchens could focus on different culinary needs, one focused around the range and the other around baking, even though the Franklin County CDC Food Processing Center we visited previously had eliminated their baking kitchen due to lack of use. Without the ability to rent two spaces at once, Brian was certain we would not be able to generate enough income to survive. As board president, it was now my job to report back to CEDF and gain their approval for an increase in funding for the construction that would be needed to follow Brian's two-kitchen proposal, as opposed to the original, uninformed proposal of just one kitchen.

As hinted at in the last chapter by my intuitive discomfort with the CEDF funding, it was when we asked to make this change that we really began to see how little power we had over the disbursement of funds. In fact, the CEO who had originally had nothing but supportive things to say about CLiCK ended up, to our shock, declaring to me and the board member who had put up the funds for the building, that CEDF should never have funded us in the first place as $100,000 would not be enough money for us to succeed. At this point, we realized that giving us the funding had been primarily in their interest, not ours as we had at first perceived. CEDF is funded by the state based on how much funding they in turn give away to small businesses. Giving us $100,000 of Department of Economic and Community Development (DECD) funds was therefore a bonus to their bottom line as well; more critically, if we failed, CEDF would have both succeeded in giving their funding away and become eligible in the end to have access to our assets, such as the building and the equipment. Hence, our anticipated failure was essentially a 'win-win' for them. Upon hearing this new narrative, the board member who had put up the funds for the mortgage – based partially on the previous positive opinion – was now quite understandably outraged and felt deceived. At the same time, having shared this disparaging opinion with us, CEDF could further use it to leverage control over the funding and how we spent it, given that we were no longer "the best idea" (but perhaps still 'a good idea', at least in theory!). As such, they opposed Brian's proposal for two kitchens, arguing that we didn't have sufficient funds and should therefore start with just one. Given that they had already told us we didn't have enough funds for such a project regardless of what we did or didn't do, I felt trapped between two scenarios both ending in failure. The first scenario was to do as CEDF desired by building only one kitchen, thereby ensuring our future economic failure as laid out by Brian. The second scenario was to push to spend the funding on two kitchens, but then there would not be enough remaining funding for operating expenses before opening, thereby assuring another form of economic failure as laid out by CEDF. Needless to say, I now found myself more anxious than ever, for not only had I, as CLiCK's president, signed for a combined amount of $270,000 (grant funding and the mortgage), I had now basically been told by various experts in the field that however you looked at it, CLiCK was destined to fail. Despite this, Tina – now in the role of our interim manager – was not anxious; instead, she was angry, determined not to let CEDF or anyone else endanger what we had worked so hard to achieve.

In reflecting upon this situation, I am not surprised to find that the website *Fast Company* notes that "…one of the biggest…buckets of despair…" for non-profits occurs when "…an organization bankrolling positive change accidentally torpedoes it by changing their funding strategy mid-grant cycle, delaying disbursement of funds for some reason, or being generally inflexible in how they dole out money that generally cash-strapped groups depend on" (Paynter 2018). A study by the Open Road Alliance found that these "…funder-created obstacles…" shockingly "…accounted for 46% of nonprofit difficulties…", suggesting "…that the biggest barrier to effective impact and the greatest pain point for nonprofits and social enterprises are their own funders" (Paynter 2018). This was a lesson we were already beginning to learn, as our funders – who had originally praised, supported, and funded us – had become our very own 'funder-created obstacle'! Obviously, the balance of power lies with the funder in these negotiations, as we didn't "…want to express dislike for some terms, lest …[we]… upset those doling out the cash" (Paynter 2018) but, we were not happy! Yet, at the same time, we still had our optimism, our own spiritual convictions, and our commitments to social justice that ultimately helped us tip the scales back in our favor.

> *I work with many start-up non-profits and have learned that vision, commitment, and enthusiasm cannot sustain a strong program – that the need for better "business" training for non-profits is essential. CLiCK underscored that concern for me – when the startup team first explored signing a lease with no real source of funds to pay the rent, for example. I had recommended starting by sharing space with an existing non-profit that had a licensed kitchen, but that raised other challenges – and the purchase of, and ongoing payment for, the current location has been reliant on the kindness of the owner and the support of the community.*
> Middle-class white female volunteer and legal advisor

Institutionalizing from without

As mentioned above, while I was internally anxious, Tina was more externally angry about our funder-created obstacle. Her anger spilled out, as mentioned, into conflict on the board, as tensions grew between those who wanted to do as CEDF dictated and those (including Tina and myself) who wanted to listen to Brian. As the board president (in my case) and CLiCK's interim manager (in Tina's case), we felt duty bound to listen to the kitchen expert, despite the increased cost, whereas those who were more concerned with the funding, including the board member who had funded our mortgage, preferred to acquiesce to CEDF. As for Tina's and my differing emotional reactions, surely part of this had to do with our personalities, but I would also argue that our evolved personalities are inseparable from our intersectional identities, and that this is in fact true for everyone. As mentioned in the Introduction, I was raised solidly middle-class, whereas Tina was raised working-class, was a young single-mother who didn't go to college, and has since her teens worked in many management roles. Tina's reaction to CEDF came out of years of hands-on negotiations in the business world, where she had learned that if you want something you have to fight for it through direct verbal engagement, backed by direct and explicit actions. In my middle-class British upbringing, on the other hand, nothing was ever addressed directly and thus actions were indirect and unclear. As our differences in addressing the conflicts around CLiCK's funding become more apparent, I was reminded of the British sociologist Basil Bernstein who focused his research on socio-linguistics, the connection between manners of speaking (grammar, syntax, sentence structure, word choice, etc.) and social organization. In *Class, Codes, and Control* (1971), Bernstein argues that not only does social class shape language use but it does so according to codes that he referred to as "elaborate" and "restricted" (p. 106). Elaborate codes are full of details, do not rely on shared knowledge, and contain more complex sentence structures, whereas restricted codes are based on shared knowledge and context and seek to serve more utilitarian purposes. Bernstein strongly argues that one code is not better than the other and each "...possesses its own aesthetic" (Bernstein 1971, p. 106). Furthermore, Bernstein affirms that the working-class uses mainly the restricted codes and the middle/upper classes use both codes depending on the situation. These class-based codes unsurprisingly have significant implications in relation to schooling and pedagogy (Mutekwe 2016), as well as how they intersect with gender and race (Morais et al. 1991). In the case of CLiCK, my reaction to the CEDF challenge was to anxiously develop 'elaborately coded' conversations in my head arguing why and how they should fund the two-kitchen scenario. Tina, on the other hand, went straight to a 'restricted code' context-based manager style retort to CEDF's representative, emphatically saying something along the lines of, "There is no way you are going to dictate how we spend the grant funds that you have awarded us", resulting in a speechless white male representative who from then on feared and avoided her. And yet, Tina's restrictive code approach resulted in the outcome we wanted, as CEDF finally agreed to reallocate funding for two kitchens, although they continued to delay funding right up to the very last payment.

What I learned from this experience and from many more that followed in relation to watching Tina assertively and astutely navigate the world of construction and contractors (many of whom engaged in all manner of unprofessionalism, including sexism as she was a female in a more male dominant profession) was that my class privilege had made me ironically *useless* when it came to actually getting complex work projects done. In the world of elaborate codes, ideas spun in discourse can float above the material world and have endless intellectual and philosophical possibilities, as in, to quote again from T. S. Eliot's *Love Song of J. Alfred Prufrock* (1917), his question as to whether or not he, "...dare to eat a peach" (p. 16). For in the end an answer must be given, a peach must be eaten (lest it rot), and reality must and will *change*. But as much as I had been trained to question and to theorize what could be, which was instrumental in the visioning aspect of CLiCK, ultimately, if CLiCK were to ever materialize as a real functioning kitchen, someone (in this case, Tina) had to make concrete final decisions, such as what material would be used on the new kitchen walls (Fiberglass Reinforced Panels), how much we would need (250 ft), how much it would cost ($32 for 4' x 8' sheet), which contractor would do it (a kitchen and bath professional), and when it would be done (ASAP!). Consequently, the dynamic between us significantly shifted, with Tina taking charge of making things happen, while the board and I tried to keep up with all that needed to get done in response. And so, construction went on for a year, along with our ever-growing concern that we didn't have enough funding to actually run the place once the construction was done. Hence, as a board, we turned our focus to the ubiquitous bane of all non-profits' existence – *fundraising*.

We had begun fundraising even before buying the building. In fact, it was, as mentioned, at an early outdoor music and wine event that the CEO of CEDF had declared to an assembly of mostly white and middle-class supporters (many of them our friends) that CLiCK was the 'best idea she had seen in years' and that people should get out their "checkbooks", resulting in our first $5,000 of donations. However, here was another area to which my social class and cultural background was ill-suited: the role of asking for money. I had and still have what Stephen R.

Block, in his book *Why Non-Profits Fail: Overcoming Founder's Syndrome, Fundphobia, and Other Obstacles to Success* (2004), refers to as 'fundphobia': "...the fear of asking people to make financial contributions to a non-profit organization" (p. 121). Despite this common fear, Block (2016) notes, "Like it or not, fundraising is a necessary activity of almost all nonprofit organizations" (p. 121), resulting in "...an estimated $427.71 billion [given]...to U.S. charities in 2018" (Giving USA 2019). Still, I was not comfortable in that role, and so I put my attention toward indirect requests such as organizing a 'Go Fund Me' (we made a fun video and raised about $4,000) and writing grants, along with another UCONN professor from the department of nutrition who had joined our board and who was also British. And he also suffered/suffers from fundphobia, and so together, we focused on grant writing instead.

Grant writing in general poses an interesting balance between Bernstein's elaborate and restricted codes; there is the elaborate storytelling narrative that spins the vision and then the restrictive nuts and bolts of how the vision will be achieved and how the requested funds will be spent. I am good at the first part and terrible at the latter, but others – including the other professor – were good at both, and so collectively over the years, we have brought in a number of significant grants from the United States Department of Agriculture (USDA) totaling about $250,000, as well as from the Connecticut Department of Agriculture (CTDOA) totaling about $100,000, plus funds from private sources totaling about $300,000. These funds eased our fears about how we would manage beyond the original DECD funds and helped us purchase our kitchen equipment and pay for an official manager (Tina was an interim one during construction) once we opened; eventually, these funds also allowed us to build a teaching kitchen and develop different forms of programming.

The federal and state funds, which ultimately come from taxes, have historically not been disbursed equally, as evidenced in the *Pigford v. Glickman* (1999) class action lawsuit against the USDA for racial discrimination against African American farmers between 1981 and 1996. Of course, racism in relation to land and agriculture goes back to the founding of this country, as mentioned, including pushing Native Americans off their lands (Kromm 2010) as part of what constitutes genocide (Madley 2015). Fortunately, after years of struggle, the *Pigford v. Glickman* case was finally settled in the plaintiffs' favor with "...a $2.3 billion settlement – the largest civil rights settlement in history" (Kromm 2010), resulting in about $50,000 per farmer. Therefore, in receiving USDA funds – recognizing our class and racial privilege as applicants – we were committed to not perpetuating their politics but rather using the funds to "...develop[...] a real community-based economic system that redistributes wealth and allows people to gain access to what they need" (Guilloud and Cordery 2007, p. 111), and to disrupt, even if in a small way, "...the neo-liberal economic model based on the increasing *privatization* of the commons" (Godfrey 2017a, p. 152). As I explored in a chapter on CLiCK for Agyeman, Matthews, and Sobel's *Food Trucks, Cultural Identity, and Social Change* (2017), although under CLiCK's model each culinary business is individually owned, "...they are mutually dependent on the success of each other in that the financial well-being of all is what will enable CLiCK to sustain itself for the collective benefit" (p. 152). Consequently, CLiCK's model has sought to "...foster *solidarity*, as opposed to competition", thereby supporting "...Elinor Ostrom's revisioning of Garrett Hardin's (1968) concept of the 'tragedy of the commons'", regarding the use and/or abuse of natural resources (Godfrey 2017a, p. 152). For Hardin, the tragedy is that users of natural resources compete with each other to maximize their personal returns, as opposed to collective benefits, thereby overusing/exploiting the resources. For Ostrom, Hardin's scenario was just one *possibility* (Ostrom 1990, p. 15), in that there exist communities that do successfully manage their commons "...from the bottom up..." to ensure "...a sustainable, shared management of resources, as well as one that is efficient from an economical point of view" (Felice and Vatiero 2012, para. 8). As such, as grant funds came in, helping us to further shape and create CLiCK, we increasingly visualized the kitchen as representing part of 'the social and economic commons', much like "the Wikipedia community" (Felice and Vatiero 2012, para. 8) or the cooperative model (Zitcer 2017) like the Co-op, or, as discussed, "commons as praxis" (White 2018). In fact, we reworked the old adage about teaching a man to fish, instead making it, '*create* a shared-use commercial kitchen and we will all eat together'! And so, despite all our struggles over the previous six years and all my fears both imaginary and real, in the spring of 2015, CLiCK finally opened its doors to much community fanfare and celebration from locals and political leaders alike. But once again, despite feeling accomplished, we had only just begun; many more changes and challenges were to come.

> *To be very honest, I put a great deal into helping CLICK become a reality and I am very glad I invested in that project. On the flip side, the process could have been streamlined a bit and we lost some good people along the way. I am thankful that I had the opportunity to run my own business, even for a short time, as I found courage and stamina that I did not know I had. My personal perspective is that I was viewed as an "outsider" of sorts and not in the inner circle. No one on the board really ever followed up with me while I ran the business and they seemed fine that I moved on.*
>
> Middle-class white female, former kitchen member

Notes

1 https://www.ohquotes.com/quote/maya-angelou-ask-want-prepared-get?bg=p54
2 https://www.permaculture.co.uk/articles/what-permaculture-part-1-ethics
3 Designed in part by UCONN students, an art professor from Western Connecticut State University and funded by a CT Department of the Arts Grant.
4 https://www.linkedin.com/in/brian-norder-829a4b23

References

Bernstein, B. 1971. *Class, code and controls.* Vol. 1. New York and London: Routledge.
Block, S. R. 2004. *Why non-profits fail: Overcoming founder's syndrome, Fundphobia, and other obstacles to success.*

Block, S. R. 2016. Founder's Syndrome in Nonprofit Organizations. *Global Encyclopedia of Public Administration, Public Policy, and Governance*. Available at: https://link.springer.com/content/pdf/10.1007%2F978-3-319-31816-5_2597-1.pdf

Butler, O. 1993. *Parable of the sower*. New York: Grand Central Publishing.

Czank, J. M. 2012. 'On the origin of species-being: Marx redefined', *Rethinking Marxism* 24(2): 316–323.

Dyer-Witheford, N. 2004. 'Species-being', *Resurgent Constellations* 11(1): 476–491.

Eliot, T. S. 1917. *Prufrock and other observations*. London: The Egoist, Fisher Rare Book Library. Available at: https://rpo.library.utoronto.ca/poems/love-song-j-alfred-prufrock [Accessed 20 February 2021].

Felice, F. and Vatiero, M. 2012. 'Elinor Ostrom and the solution to the tragedy of the common's [online].

Franklin, A. 2002. *The sacred circle tarot: A celtic pagan journey*. St. Paul, MN: Llewellyn Publications.

Fromm, E. 1974. *Marx's concept of man*. New York: Frederick Ungar Publishing.

Giving USA. 2019. 'The annual report on philanthropy for the year 2018' [online]. Available at: https://givingusa.org/giving-usa-2019-americans-gave-427-71-billion-to-charity-in-2018-amid-complex-year-for-charitable-giving/ [Accessed September 28, 2020]

Godfrey, P. 2017. 'Reflective food truck justice: A case study in CLiCK, Inc., a nonprofit, shared-use commercial kitchen'. In Agyeman, J., Mathews, C. and Sobel, H., eds. *Food truck, cultural identity and social justice: From Loncheras to Lobsta love*. Boston, MA: MIT Press, pp. 149–167.

Godfrey, P. 2017a. 'Radical pedagogical homesteading: Returning the 'species' to our ' being''. In Haltinner, K. and Hormel, L., eds. *Teaching economic inequality and capitalism in contemporary America*. New York: Springer, pp. 91–103.

Guilloud, S. and Cordery, W. 2007. 'Fundraising is not a dirty word: Community-based economic strategies for the long haul'. In INCITE! Women of Color against Violence. *The revolution will not be funded: Beyond the non-profit industrial complex*. Cambridge, MA: South End Press, pp. 107–113. Available at: https://newriverabortionfund.org/wp-content/uploads/2020/07/the-revolution-will-not-be-funded-beyond-the-nonprofit-industrial-complex.pdf [Accessed 20 February 2021].

Hardin, G. 1968. 'The tragedy of the commons', *Science* 162(1): 1243–1248.

Hewitt, B. 2009. *The town that food saved: How one community found vitality in local food*. New York: Rodale Books.

Kromm, C. 2010. 'The real story of racism at the USDA: The USDA's real race problem is its history of discrimination against African-American, Native American and other minority farmers who were pushed off their land' [online]. *The Nation*, July 23. Available at: https://www.thenation.com/article/archive/real-story-racism-usda/ [Accessed September 28, 2020]

Madley, D. 2015. 'Reexamining the American genocide debate: Meaning, historiography, and new methods', *American Historical Review*. February, 2015.

Marx, K. 1844. 'Economic and philosophical manuscripts of 1844'. *Estranged labor*. Available at: https://www.marxists.org/archive/marx/works/1844/manuscripts/labour.htm [Accessed September 28, 2020]

Miles, M. 2010. 'The art and science of making a huglekultur bed-transforming woody debris into a garden resource'. [online] *Permaculture news*. Available at: https://www.permaculturenews.org/2010/08/03/the-art-and-science-of-making-a-hugelkultur-bed-transforming-woody-debris-into-a-garden-resource/ [Accessed September 28, 2020]

Morais, A. M., Peneda, D. and Madeiros, A. 1991. 'The recontextualizing of pedagogic discourse: Influence of differential pedagogic practices on students' achievements as mediated by class, gender and race', *International Sociology of Education Conference*. Birmingham: University of Birmingham.

Mutekwe, E. 2016. 'Interrogating the social class assumptions and classroom implications of Bernstein's pedagogic discourse of visible and invisible pedagogies', *Journal of Sociology and Social Anthropology* 7(2): 118–125.

Ostrom, E. 1990. *Governing the commons: The evolution of institutions for collective action*. Cambridge: Cambridge University Press.

Paynter, B. 2018. 'How funders often hurt the nonprofits they are trying to help' [online]. *Fast Company*. Available at: https://www.fastcompany.com/40561244/how-funders-often-hurt-the-nonprofits-they-are-trying-to-help [Accessed September 28, 2020]

White, M. M. 2018. *Freedom farmers: Agriculture resistance and the black freedom movement*. Chapel Hill: The University of North Carolina Press.

Zitcer, A. 2017. 'Collective purchase: Food cooperatives and their pursuit of justice'. In Alkon, A. and Guthman, J., eds. *The new food activism: Opposition, cooperation, and collective action*. San Francisco: University of California Press, pp. 181–205.

Part III

Thresholds of successes, failures, and unknowns – explorations in *praxis*

7 Putting just sustainabilities and intersectionality into *praxis?*

An ounce of action is worth a ton of theory.

Ralph Waldo Emerson

"Real isn't how you are made," said the Skin Horse. "It's a thing that happens to you. When a child loves you for a long, long time not just to play with, but RE-ALLY loves you, then you become Real."

The Velveteen Rabbit

Who or what is 'community'? How can we create a just community?

The last chapter covered the purchase, build-out, and grand opening of CLiCK's (Commercially Licensed Co-operative Kitchen's) building as a shared-use commercial kitchen (the teaching kitchen was not funded or finished until the following year). These actions happened over five years ago, took more than a year and a half to complete, and still did not (do not) ensure our success. Although after construction we physically existed, we had not yet fully 'become *Real*' beyond our own small board and corresponding social networks. What we needed, much like the conversation in the Velveteen Rabbit, between him and the Skin Horse (Williams 1995, p. 14), was to gain the collective *love* and input of the larger cross-cultural *community* through targeted actions, open solicitations, grand visions, and unrelenting dedication in order to gain kitchen members and community support. Doing so seemed to me to be the only way we would succeed, keeping the momentum going and channeling my ongoing anxiety into action. Additionally, as in the previous discussion of Cidell's (2017) concept of the 'sustainable imaginary' (2017), we wanted to welcome kitchen and community members into the envisioning process of creating our 'JS imaginary', while recognizing that it was, as Emerson notes, the 'ounce of practice'[1] that would matter in the end.

Yet, it is worth pausing a moment here to unpack the term 'community' and its implications. David M. Chavis and Kien Lee (2015), writing for the Stanford Social Innovation Review, note that, "'Community' is so easy to say .... It seems so simple, so natural, and so human", especially as the term is often used "...as a symbol of good intentions" (para. 1). However, they go on to say, "...the meaning of community is complex", in that it is "...both a feeling and a set of relationships among people...", and involves the delineation of "...who is and isn't part of [our] communities" (para. 4), as well as how our memberships in different communities can overlap, intersect, and even contradict. Additionally, communities involve a sense of place, as well as formal and informal institutions operating according to specific temporal social norms that are collectively enforced in varying ways, to varying degrees, and at varying times. As such, when discussing communities, applying intersectionality as an analytic tool is helpful to deconstruct the implied 'simplicity' and aid in revealing how communities manifest the "...complexity in the world, in people and in human experiences" (Collins and Blige 2016, p. 2). This is especially relevant given the role of community in knowledge production, as Lynn Nelson (1990) argues that "...knowledge..." is collective and "... the primary epistemological agents are groups – or more accurately, epistemological communities" (p. 256), making what we know inseparable from who we are, collectively. Of course, communities are *intersectional*, they are as Dara Cooper observes in her Foreword to Ashanté M. Reese's *Black Food Geographies: Race, Self-Reliance, and Food Access in Washington, D.C.*, "...nuanced ...never monolithic" (2020, p. xiii). However, as she goes on to add "...a direct by-product of anti-Blackness...", and I would add other forms of racism, is that "... mainstream dominant narratives rarely afford Black communities the dignity or humanity of nuance" (p. xiii). Instead, Black Indigenous People of Color (BIPOC) are "... represented in flat, one-dimensional narratives that often have the subtheme of nothingness or lack running throughout" (p. xiii). In contrast, when communities are analyzed intersectionally, it reveals how they have been differently constructed, valued, oppressed, and/or empowered, both within themselves and in relation to other communities, depending on the intersecting and unequal social identities of their members and the outside 'others'.

Given that CLiCK was started and run mostly by middle-class whites, typical of non-profit boards (Thurman 2011), creating an inclusive and cross-cultural, nuanced community required recognizing the role of our privileges and actively addressing the intersections of sexism, racism, and classism as best we could within ourselves, our organization, and the larger society. This has not been easy, hence the popular term and practice of 'difficult dialogues'[2] and has at times been a point of contention on our Board, when the issues of structural inequalities were more overtly brought up by me, Tina, and/or other more progressive board members. As Robin DiAngelo, in her book *White Fragility: Why It's So Hard for White People to Talk About Racism* (2018), notes, whites like us have been, "Socialized into a deeply internalized sense of superiority that we either are unaware of or can never admit to ourselves..." and so "...become highly fragile in conversations about race" (p. 2). This point rings true from my experiences as a white person in relation to myself, in teaching white students about white racism, as well as in particular in my experiences with white people at CLiCK. However, it is important to note that DiAngelo's work has been rightly critiqued for espousing white "homogeneity over difference" (Bejan 2020, para. 6), thereby erasing other intersecting social divisions such as social class and gender. Additionally, as Columbia University Professor and writer at *The Atlantic* John McWhorter (2020) ironically asserts, DiAngelo's book is "…a racist tract…" that "… diminishes Black people in the name of dignifying us" (para. 3). According to McWhorter, DiAngelo does this in her assertion that "…all disparities between white and Black people are due to racism of some kind" (para. 12) – a racism that whites can apparently never fully overcome, thereby ensuring their ongoing power over seemingly powerless Blacks. Obviously, I agree that DiAngelo lacks a critical intersectional analysis and also, as McWhorter identifies, that her work takes on a "cult"-like tone, making all whites into sinners who must atone, but can ultimately never be fully saved (para. 11). Nevertheless, as uncomfortable as I am with promoting work that in this case McWhorter, a middle-class person of color (he speaks to his class privilege but not that of his gender), finds 'racist', I still think, based on my own experiences, that her insights are helpful in *moderation*. I say this in that they invite *us* whites to be the ones who must address racism within our organizations and ourselves, especially if we identify as white and progressive, as I and some others on our board did/do. For based on DiAngelo's years of research and experiences as an anti-racist educator, she has found that despite popular perceptions around white racism and social class, as in the notion that the 'real racists are poor whites' (an example where she does look at class but does not engage in an intersectional analysis), "…white progressives cause the most daily damage to people of color…" through erroneously believing and insisting that we have already "…arrived" (p. 5). This insight echoes what Malcolm X once said about 'white liberals' being much more dangerous than 'conservatives', as in being devious like a 'fox' who is able to "…lull you with his smile" (quoted in Duncker 2020). And if I am honest, I must admit that I have seen evidence of this claim of having 'arrived', of being like 'foxes' in *myself* and other whites involved in the Alternative Food Movement (AFM) and even the Food Justice Movement (FJM), as will be discussed further in the next chapter. Suffice it to say here that our approach as CLiCK's board was more aligned with trying to uphold the seven cooperative values, as well as trying to create an inclusive environment by putting into practice what Martin Dempsey and Ori Brafman (2018) have termed "…radical inclusion", rather than directly engaging in anti-racist work, as also evident in their book written more for corporate leadership than social justice work. Nevertheless, their term resonates with me even as our practice fell short of overtly and *actively* addressing structural inequalities within CLiCK and in the wider food system. At the same time, I would assert that we *tacitly* (as opposed to overtly) recognized structural racism in our local community and within the food system as a whole both historically and currently, as I will later discuss in relation to our failures.

Given CLiCK's non-profit, cooperative mission as a small shared-use commercial kitchen, we sought to practice 'radical inclusion' by delivering economic opportunities for all within the culinary field (as well as eventually nutrition and culinary education), across differences of social class, race, and country of origin, including the local immigrant community hailing mostly from Puerto Rico (again recognizing that Puerto Ricans are American citizens), Mexico, as well as Central America.[3] By supporting the incubation of small-scale food businesses for those with limited capital investment, our goal was to enact progressive social change, hence situated social justice (Aldarondo 2013, p. 152). We wanted CLiCK to not only be 'a symbol of good intentions' but, as mentioned, to share our resources *with* others (McLaren and Agyeman 2015), creating, as explored, "commons as praxis" (White 2018). Yet, in all this arose the larger questions of what the actual creation, maintenance, and sustainability of 'justice' could look and feel like at CLiCK, as well as the questions as to what a 'community' that could *transcend* the normalized structural boundaries of race, gender, and class would look like and how could we achieve it '*with*' others, rather than *for* 'others' or even *without* 'others'. In short, we began to wonder how could CLiCK continue to 'become Real', hence *loved*, by those from differing 'epistemological communities'? Additionally, we wondered how could CLiCK 'become Real' when the shared-use non-profit cooperative model required co-construction by others, creating their own businesses and or events within our creation and thereby becoming real themselves? To explore how we answered these questions, I will unpack in more details what I have come to understand as the 'non-profit conundrum', as well as the ways we attempted to create a 'radically inclusive organization/community' based around food that we hoped would, for different people in different ways, embody some aspects of the just sustainabilities (JS) imaginary.

My involvement with CLiCK was filled with mixed emotions. As much as I wanted to contribute to CLiCK's success as a board and as an organization, my way of doing things seemed so contrary to what other board members desired that it became very difficult and at times very discouraging to be a board member. Of course, any experience in one's life is a learning experience and I am grateful for meeting all of the great people that were on the board while I was there. They demonstrated serious commitment and support for CLiCK and that was really nice to see.

Middle-class Latina former board member

Challenges to *just* institutionalizing

The most obvious first step in answering the questions posed in the previous section (all of which link to the larger questions of this book), now that we had United States Department of Agriculture (USDA) grant funds to hire a part-time general manager (GM), was to prioritize hiring someone from within the Latinx community to ensure representation and to take our commitment to equity seriously. As such, we posted our job description strongly encouraging bilingual applicants and instituted the federal government's equal-opportunity hiring guidelines. We established our criteria rubric and had three board members interview potential candidates, including several members of the Latinx community, as well as several veterans; using the gathered data and feedback, we then voted on the candidates. We ended up choosing a non-bilingual white college-educated female with a background in the culinary arts and a focus on childhood nutrition. It could therefore be argued that despite 'our talk' to valuing diversity, we had failed our first significant challenge in 'our walk'. However, in hiring a white GM, CLiCK was of course not alone. The Common Good Careers and Level Playing Field Institute[4] report from 2011 found that "...nonprofit employees are approximately 82% white, 10% African-American, 5% Hispanic/Latino, 3% other, and 1% Asian or Pacific Islander" (Thurman 2011). Writing about this study, Rosetta Thurman (2011), a consultant in non-profit management, leadership, and social change, asserts that the reason this racial disparity continues must be that non-profits "...don't really care" (para. 12) about addressing structural racism within their organizations. Helen Kim Ho (2017) also notes the frequency of "...tokenism", in which people of color are used "...for their colorful (pun intended) personal stories of hardship and discrimination to pull at the heartstrings of donors or legitimize the cause", but when they "... try to exercise their leadership roles, those [whites] in power will work to undermine, block and derail them" (para. 7). As tempted as I am to add an explanation or an excuse as to why CLiCK hired who we did, I will not. I will however say that, in my experience, 'caring' about an issue does not guarantee that one is able to create the desired outcomes at the time, place, or in the desired manner. As for outright racism or implicit bias in our hiring, it is impossible to separate our individual actions from those of the larger society wherein racism plays an incipient role in all aspects, shaping our identities, opportunities, and communities (Bonilla-Silva 2007; DiAngelo 2018; Omi and Winant 1994). Therefore, addressing structural *and* interpersonal racism must become an ongoing topic of critical analysis and engagement throughout white society, including in our non-profits, which for the most part are *not*, according to Thurman, "... walking that warm and fuzzy 'everyone is welcome' talk" (Thurman 2011), at least not when it comes to sharing power.

Having imperfectly hired our first GM, we were still nevertheless committed to actualizing our social justice mission by making the totally overwhelming, unachievable, and underpaid GM position as tolerable as possible. In addition to the lack of non-profit employee diversity, another common area of contradiction in relation to non-profit missions stems from the exploitation of their employees, particularly with regard to gender. According to Nonprofit Quarterly, 73% of non-profit employees are female (and as discussed from an intersectional perspective are also white and no doubt more middle-class, but I could not find data supporting this), although non-profits with larger budgets are more likely to have more males on their "...executive board and staff leadership" (Shapiro 2019). One proposed reason is that under the inequalities of patriarchy women tend to be "...more mission driven" (Shapiro 2019), as opposed to career driven and that women – in particular white middle-class women given their history of volunteerism, including volunteer tourism to 'save' those less fortunate in the global south (Bandyopadhay and Patil 2017) – can afford to work for lower wages in exchange for what they perceive as more meaningful employment (McCarthy 2001). Also, straight women who are married to men can often rely on their spouse's additional and typically higher income, especially if they are middle-class. This emphasis on the 'mission' (given its religious roots) accompanied by high labor demands on the one hand and limited resources on the other causes non-profits to place an excessive physical and emotional burden on employees (Timm 2016), who are predominantly white females. Additionally, emblematic of patriarchy's misogyny is a comprehensive study of occupational feminization and pay, which demonstrates how the higher the percentage of women in a given occupation, the lower the social value and therefore the lower the wages (Levanon et al. 2009). Furthermore, this contradiction between non-profits' missions and their labor practices is another reason, according to Helen Kim Ho, that non-profits "…fail at being the most diverse, safe and woke-est place imaginable", as "…many POC simply can't afford the low pay of nonprofit life" (Kim Ho 2017). A case in point: our GM was married when hired, but two years later, when her marital status changed, she left the position for a full-time one that would enable her to support herself and her child. Thus, the challenges non-profits face if they do want to 'care' about social justice in their actual day-to-day operations are deeply rooted and extensive within the larger society and not just within the confines of their single organization. As such, social justice as a practice requires that we at CLiCK and elsewhere continually try to face "…the contradictions and complexity of everyday life" (DuPuis et al. 2011, p. 298) and to focus on our "…process", while still not allowing the 'perfect to be the enemy of the good', in order to nevertheless commit to making the 'good' better, more inclusive, and ideally more just.

We sought to mitigate the burdens written into our GM position by continuing to act as a working board and supporting her to the best of our ability – never acting as if we had all the 'theory' and she was to do all the 'practice'. Using private foundation funds from the Jeffrey P. Ossen Foundation, over the following year, we were able to build a teaching kitchen and to hire an additional part-time employee who would focus specifically on nutrition education, mostly in Spanish. This part-time position was created *with* and *for* a Latina who already worked as the nutritional outreach coordinator for University of Connecticut (UCONN) Extension to the Latino community and who had formerly also been on the board as our vice president. With the completion of CLiCK's teaching kitchen, she was finally able to offer her existing classes as well as new ones in an actual kitchen, enabling others to share in cooking *with* her rather than just watching her cook on a plug-in electric stove top, as was her previous practice due to her students' lack of space and transportation options (still an ongoing issue for those individuals without cars who desire to get to CLiCK). She also became invaluable in helping our then-GM and our next GM with our Spanish-speaking entrepreneurs using CLiCK's commercial kitchen, some of whom have started and ended businesses, while still others have had great success, as will be further discussed in the next chapter.

Another way that CLiCK would increasingly become Real to the larger community was to have our new GM focus on recruiting culinary entrepreneurs, supportive individuals, and/or institutions to become members of CLiCK, as outlined in our by-laws and as part of our cooperative identity. Her goal was to create a public narrative about CLiCK that would attract members, who would in turn help supplement the grants and private funds that had enabled us to get to where we were but could not cover daily operations. A significant challenge was finding a balance between making CLiCK affordable to culinary entrepreneurs while also being able to meet the costs of running two commercial kitchens. Although non-profits don't pay property taxes, they are still subject to the tenets of capitalism, paying the market price for electricity,[5] oil, and other necessities, all of which in our case would increase the more the kitchen was used. Thus, like most grassroots non-profits, CLiCK has struggled financially since its inception, in part because funders prefer to fund specific programs as opposed to the nuts and bolts of running an organization.[6] Moreover, many of the individuals and institutions whom we hoped to encourage to become paying members were not entrepreneurs (including other non-profits, schools, universities, neighbors, etc.), and so, we had to offer them a membership experience that did not rely upon business incubation. To this end, we sought to formalize an array of food-based free events designed to be the mechanisms by which individuals would develop feelings and relationships connected with CLiCK and each other, making the CLiCK idea and community a 'part of them', as will be discussed in the next section. This focus included, as previously mentioned, trying to create appropriate outreach materials for our food events that were accurately translated into Spanish by members of specific Latinx communities, thereby reflecting that community's use of Spanish, including any appropriate slang and or imagery to avoid homogenizing distinct cultures.

As I have explored previously, the first culinary members that came to CLiCK were mostly "…white nationals…from the middle class and had social and economic resources to start their businesses" (Godfrey 2017, p. 154). As a result, given our mission, the board and the GM reevaluated our rates for kitchen use and ended up reducing the rental rates, thereby engaging in an albeit "imperfect" example of "reflexive food justice" (DuPuis et al. 2011; Godfrey 2017) in that the costs still had to be paid by CLiCK, but we attempted to shift the burden from individual members to the collective organization. Melanie DuPuis, Jill Harrison, and David Goodman in their article "Just Food?" (2011) propose the concept of "reflexive food justice" that acknowledges "…the tensions between different definitions of justice, the environment and the bodily health, and good food, while admitting that local strategies are imperfect and contradictory" – hence "imperfect" (p. 297). As such, they proposed seven imperfect definitions which I have used in the past in relation to CLiCK (Godfrey 2017) and they all still apply. In particular, the first one has become increasingly apparent in this larger analysis of CLiCK's story, which is the acknowledgment that "Reflexivity begins by admitting the contradictions and complexity of everyday life" (p. 297); a perspicacious statement that becomes increasingly valuable when trying to engage in progressive social change.

Reducing the cost to use CLiCK's kitchens made them more economically accessible to a wider range of potential entrepreneurs, ultimately increasing the social class/racial/gender diversity of our members – over half of our businesses have since then been female-owned, about 1/3 have been minority-owned, and all have been started with minimal capital and no more than 1–2 employees. On the other hand, this decrease in use fees only further increased CLiCK's own financial struggles. We did consider having a sliding scale based on members' 'taxable income', but decided against this approach since many members were starting new careers and the required 'policing' to enforce this policy would be counter to the sense of community we were trying to create. Alternatively, over the years, we have secured numerous grants to specifically help low-income entrepreneurs offset their costs for membership, insurance, licensing, and supplies. These funds have been instrumental in helping us address the complex yet predictable realities of racial, gender, and economic inequalities through intentional and strategic interventions, both monetary and social. Ultimately, even though not all of CLiCK's member businesses have been successful, they have all at least been given some measure of equal economic *opportunity*.

Another challenge that CLiCK acknowledges upfront is the fact that a high percentage of food businesses/restaurants fail – as many as 60% in the first year and 80% by the 5th year (Bellini 2016). On top of these inauspicious odds, many of our entrepreneurs experienced struggles with child care, additional capital input, and lack of family support. A memorable example was the case of two women from West Africa who had met with our GM and were very excited to start their businesses, only to be denied the 'permission' to do so by their husbands. Nevertheless, CLiCK was committed to giving people the opportunity to dream and to *try* to actualize those dreams, knowing that if they did fail for whatever reason, CLiCK's affordability would ensure that they could do so without incurring excessive debt. In fact, part of our GM's job was to not only help support and oversee those starting food businesses but also to advise those who seemed excessively ill-prepared to *not* become members. This proved to be another challenging balance for our GM, as CLiCK never wanted to take people's money when it was obvious they would not succeed, while at the same time we did not want to put a damper on people's dreams.

Finally, we knew that those few who *did* achieve success, as in sustaining themselves enough to grow their businesses, would eventually move on from CLiCK in search of more space or a better location. Indeed, this has already happened, as a successful kombucha business incubated at CLiCK has since moved on, and an increasingly successfully prepared mail-order meal company may soon outgrow us. And so, given all these intersecting complexities in our business incubation model, relying on the *success* of other businesses for us to have our own *success* is a difficult line to walk. Those we seek to help the most are the least able to succeed, creating an ongoing contradiction between our mission and the external structural inequalities, while those members who do find great success may ultimately grow beyond what CLiCK can offer – resulting in a continually tenuous outlook for the organization at large. Hence, attempting to institutionalize CLiCK's shared-use kitchen in affordable and *just* ways only increased 'the contradictions and complexity' of our 'everyday' operations.

> *It gave an expanded perspective on the community in which I live. Gave me an opportunity to practice cooperative values on a daily basis, understand how a true community/cooperative organization could work. It occupied a major part of my life for the better part of 10 years, and at this time, I am happy to have moved on, but would not trade the experience.*
> Middle-class white female former board member
>
> *CLiCK is a real business kitchen…and I have more employees and support… and my house and children no longer smell always of food.*
> Middle-class Latina kitchen member and culinary teacher

Becoming *collectively* real

At this stage of CLiCK's development, our role as the board was to support the GM, to focus on the financials (our P&L – profit and loss), and to let as many people as possible know about CLiCK in order to create a 'radically inclusive' organization/ community. One of the ways Tina (now back on the Board) and I chose to do this, with the help of a few other willing board members, was by collaborating with other local non-profits, as well as *with* many volunteers from the Latinx communities, to host events focusing on *food*, specifically two celebrated Latino foods: Mexican salsa and Puerto Rican coquito. At this time, our board had grown, and new members had joined and left and joined and left (we remained majority white and female with never more than one or two Latinx members, as well as having income, age, sexuality, and ability diversity), with the frequent turnover largely due to the amount of work and the financial stress of often being on the verge of closing. An ongoing struggle was that not everyone on our board could put in the time and energy needed for all that there was to do. Additionally, my leadership style was very much shaped by my pedagogical practice as an engaged educator (hooks 1994), meaning that in my classes, I seek to use my power to empower, hence engage, others. As such, as CLiCK's president, I sought to practice a similar model of 'engaged leadership', that was open and transparent in order to be as inclusive as possible. However, many who joined the board expected to be told what to do, as opposed to being invited to do what interested them, to serve on existing committees (such as grant writing, education, fundraising…etc.), or to come up with their own ideas. As a result, instead of inspiring people, as I had imagined, in many cases, it drove people away in that, as hooks identifies, for some "…transgressing boundaries…[is] frightening" (p. 9). In fact, one of the board's more conservative members would refer to my and therefore CLiCK's lack of authoritarianism as our 'loosey-goosey', which although said more as an ongoing complaint, I nevertheless took as a compliment as it implied that there was space for us to grow and for the unknown to manifest.

At the same time that I sought to practice 'engaged leadership', I also tried to make it apparent to everyone on the board that for CLiCK to 'become Real' to people from differing communities, we needed to be constantly working on building inclusive relationships in a manner that would run counter to our highly racist and segregated society, while helping to more formalize our mission. Yes, we sought to practice 'radical inclusion', but we also worked to recognize that inclusion is *not* only a subjective feeling but must also be intentionally and structurally created. This point is confirmed by psychologist Deborah L. Plummer in her article *Getting to We: Inclusion is More than A Feeling* (2017) who states that an organization (or community) must create, "...an inclusive culture" (para. 5), which includes eight "...inclusive factors" (para. 7). Examples of these are "Common purpose...", "Trust...", "Access to opportunity...", "Sense of belonging...", and "Respect..." (para. 7) all of which, as she goes on to note, "...are interdependent and should be considered in the aggregate". As such, when all are present, "...inclusion can then be defined...as a set of social processes where individuals experience...", to offer just two: "access to information and social support; security within their identity group or in a position within the organization" (para. 8). Additionally, she identifies that key to achieving these "...inclusion dynamics..." is to reinforce and embed them into "...an organization's culture..." (para. 9), as in its:

-mission, vision, values: uses inclusive language and specifically references diversity
-strategy, structure, systems: organization is structured to allow for diverse ways of
    knowing, limits bureaucracy, and information and resources are accessible
-policies, practices, procedures: open, transparent, and consistently applied.

(para. 9)

I quote Plummer extensively here as in retrospect I find her insights highly apposite to how we attempted to create our 'radically inclusive' organization/community based loosely on the seven cooperative principles, as well as on CLiCK's mission and own social justice commitments. Additionally, her recognition of the importance of institutionalizing such a culture of inclusion in order to promote "The *feeling* of inclusion... [italics added]" (para. 11) is key, in that doing so helps to mitigate against the vagaries of human emotions, the fragility of human conversations, and the counter inclusion narratives of the larger social structure. We attempted to do this through our commitment to being membership-based, as per our by-laws, by trying to formalize the inclusiveness of CLiCK through the shared kitchen and classes, as well as seeing the role of food as being able to bring socially divided, segregated, and unequal communities together in authentic and mutually beneficial cultural celebrations.

In his brilliant book, *The Sacred Canopy: Elements of a Sociological Theory of Religion* (1990), Peter Berger argues that we create our worlds "... in conversation with others and... both identity and world remain real to ...[us] only as long as [we] can continue the conversation" (Berger 1990, p. 16). Hence, the need for institutionalization as elaborated on in his previous book *The Social Construction of Reality* with Thomas Luckman (1967) where they recognize that, "Institutions... by the very fact of their existence, control human conduct by setting up predefined patterns of conduct..." (p. 55). In other words, making CLiCK real and making that reality manifest individual and collective patterns of behavior that could then engender feelings of inclusion across differing communities, required us to not only keep the conversations going, but to also formalize the mechanisms to make them happen. As such, we sought to 'control human conduct' while not fully blocking "other possible directions" (p. 55) of growth in order to allow the tenets of our JS imaginary to act as an emergent property. By hosting events based on a celebration of culturally specific foods, we sought to formally make manifest our culture of inclusion, while allowing others to add in their own creative ideas so that the events modeled, however briefly, experiences of a more *just* and sustainable world. Such a world was not, as previously emphasized, 'imaginary' (Anderson 1983; Cidell 2017) as in made up, but rather was collectively *real* in that it was based on shared beliefs about how the world works (both in terms of how it can and should work), as well as shared experiences about how the world was working at that moment through our collective 'conversations'. Furthermore, as John Robinson and Raymond Cole provocatively observe, "…sustainability can usefully be thought of… as the emergent property of a conversation about desired futures…" (Robinson and Cole 2015, p. 137), and as such, much of what we were doing was talking about and trying to act out our 'desired futures' and thereby helping to formalize them into being collectively real. Interestingly, the more we were able to grow and share our visions, thereby making them more collectively real, the less anxious I became, as I no longer felt singularly responsible for their emergence. In fact, as in the title of Plummer's article, I began to see that we were finally "Getting to We" (2017).

Our first salsa festival was held in September 2015 and was spearheaded by a local young Mexican American Vista volunteer who worked with an emerging youth gardening organization called Grow Windham (GW).[7] GW was co-founded about the same time as CLiCK by the Willimantic Co-op's manager and a white middle-class former female teacher from a neighboring town. GW's focus has been on youth leadership and development in relation to growing a "…stronger community and local food system"[8] and they have continued to grow to now run a small urban farm in Willimantic, as well as afterschool and summer programs. In addition, GW spearheaded another food justice organization, the Windham Community Food Network, that has over the last few years been collaboratively working to, "…create opportunities for the community by building a healthy food network"[9] along with other community partners, including CLiCK.

Given our shared interests and other obvious commonalities, the co-founder and director of GW became our friend; however, one key difference was that she was married to a high earning male, enabling her to volunteer full-time, which we could not do. Likewise, eventually, the dynamic Vista volunteer could not sustain himself on the paltry salary (about $1,200 a month), a salary often justified by the assumption that since volunteers will be working with those who are poor, they too must experience poverty. Of course, as discussed, this reinforces the charity model that only those of privilege can serve; although I could not find data on the race/gender/social class make-up of AmeriCorps/Vista volunteers, I suspect that like the non-profit field, it is majority white middle-class and female. However, in

Figure 7.1 CLiCK Day of the Dead Traditional Mexican Dancing.

the year that this Vista volunteer did serve with GW, both their organization and ours were greatly enriched from his connections with the local Mexican community, highlighting the importance of hiring individuals from the communities we sought to serve.

After the collaborative success of our Mexican salsa festival (which included traditional Mexican dancing, Salsa dancing lessons taught by a teacher from Colombia, salsa tasting and judging by Willimantic's mayor and a Mexican elder, as well as much cross-cultural engagement among over 75 attendees) and Day of the Dead festival (see Figure 7.1), CLiCK gained its first Mexican immigrant member, again indicating to us the importance of social networking opportunities – something from which low-income/women of color entrepreneurs are often excluded (Barr 2015; Godfrey 2017). However, in an unforeseen outcome, our former vice president and now nutrition educator at CLiCK shared with us that in her Puerto Rican community we were now known as 'that Mexican place'. This prompted us to further acknowledge the complexity in trying to become Real across *communities* as a multicultural organization. Additionally, it challenged us to deconstruct the problematic and homogenizing terms 'Latinos', 'Hispanics', and even 'Spanish people', when in fact there are complex limits to "...panethnic cohesion", with many individuals preferring to identify by "...country of origin or ancestry" (Lee et al. 2017). Thus, she suggested that we host a coquito festival in

December, celebrating a traditional Puerto Rican holiday drink along with music, traditional dancing by a middle-school group, food, and of course much coquito tasting, as well as again including judging by our town mayor and a well-known member of the local Puerto Rican community.

This event was another cross-cultural success in that we had about 50 people from diverse segments of the community come together to enjoy and judge coquito and to celebrate Puerto Rican culture. As a result, we sought to continue the momentum of trying to balance formalizing these events while still allowing them to organically evolve. Case in point was that following spring three of my students from UCONN (all female – two white and one Afro-Latinx) completed an internship at CLiCK and took the initiative to organize a cross-cultural Earth Day event, attempting to counter the conventional whiteness of the day (Mangan N/A). We also held a summer community meal and sign painting event in the orchard for other non-profits, including churches (another cross-cultural experience, see Figure 7.2), as well as many kinds of cooking classes in English and/or Spanish, including ServSafe food safety trainings. In addition, we held art-making events, theater, and live music events (including hosting a traveling group featuring performers who were Lakota, Palestinian, and Israeli/Bedouin), and speaking events co-sponsored by UCONN with well-known leaders in the FJM, such as Malik Yakini (see Figure 7.3) founder and Executive Director (ED) of the Detroit Black Community Food Security Network, which, "…is grounded in an antiracist, anticapitalist mindset and emphasizes cooperative effort and collective wealth building" (White 2018, p. 121). We also hosted Mark Winne (see foreword), who was ED of the Hartford Food System, co-founder of the Community Food Security Coalition (no longer in existence as mentioned), and the author of numerous food movement books. All of these events were free, open to the public, and aimed at bringing diverse community members together. We intentionally sought to counter a ubiquitous mantra in our dominant white community (and no doubt elsewhere) in relation to the local 'Latino community', which goes something like, "They are always invited to community events but they don't come". And yet as discussed, 'community' is never neutral and without creating an intentional 'culture of inclusion', which includes culturally appropriate collaboration on the part of whites *with* BIPOC communities, their members will not *feel* authentically welcome. Furthermore, although such outreach was an incredible amount of physical and emotional work, we felt it was essential if we were to, as Joshua Sbicca and Justin Sean Myers (2015) state, "…reshape racial [and class] identities, meanings and structures…" even if only in micro interpersonal and organizational ways. Furthermore, by trying to practice radical inclusion *within* CLiCK, we sought to have it seep out into the larger community, even as the dominant counter values of broader society nevertheless continued to taint us.

Other ways that we attempted to practice radical inclusion was by doing outreach at summer street festivals, farmers markets, and health fairs, and I, as president, continually got myself invited and others (staff and/or members) onto local radio and television shows; I also wrote articles about CLiCK for a local paper called *Neighbors*[10] and tried to get other press coverage. However, none of these outreach efforts raised any significant funds, and so on top of these events, we also organized two major in-person fundraisers a year (along with solicitation by

Figure 7.2 Community Meal and Sign Painting in Orchard.

Figure 7.3 Malik Yakini, Holding a Workshop.

mail), which featured the foods of our culinary members and were based on ticket sales that included such things as local wine tastings, music shows, films, garden parties, cooking classes, etc. All of these were attended overwhelmingly by other white middle-class individuals, including many people we already knew.

Another way we did fundraising was inspired by Mark Winne (who at the time we had not met), titled "The Fundraising Letter I'd Like to Receive", in which he highlights the tendency of food banks to keep harping on the crisis, hence our heartstrings, as opposed to offering ways they are helping to create solutions. He goes so far as to write his ideal letter that ends stating, "So please help us continue our work of empowering more and more of our clients and neighbors so that we can distribute less food year after year" (Winne 2013, para. 14). Using this as a model, we wrote similar fundraising letters inviting our community to help CLiCK support members starting food businesses, taking nutrition classes, and doing other things that were aimed at helping people change their circumstances (Figure 7.4). As such, we emphasized that,

> ...CLiCK isn't your typical non-profit. We are in fact on the cutting edge of being a new model by helping others help themselves .... – while remaining true to our vision for a more just, healthy and sustainable food system, both locally and globally.

> (Godfrey 2015)

It would be impossible to tell if such an appeal was officially more or less effective than the usual charity format critiqued by Winne, but I felt such an approach fits better with our social justice, as opposed to charity values. As such, there were those that did appreciate what we were trying to do, although for others, it was apparent that the charity model was an easier sell. Raising funds by emphasizing the needs of our low-income entrepreneurs through an empowerment approach further helped to address the division in priorities between the free cross-cultural community events and the fundraising events targeted at the white dominant community. This division was continually played out in conflicts on our board between our more conservative and fiscally realistic members and our more social justice–oriented and fiscally idealistic ones (Godfrey 2017, p. 162), continuing the previously discussed division between the tenets of the AFM and the FJM.

Similar conflicts to those on our board arose with other non-profits, including our community events partner GW, in terms of who was doing the actual outreach work and who was unequally benefiting, resulting in a temporary dissolution of our formal relationship. In addition, as will be further discussed, these conflicts illustrate how interpersonal and organizational relationships are of course pervious to the larger unequal and oppressive social dynamics of race, gender, and class, even when these relationships take place between members of the *same* race, gender, and class. This observation can be more obvious in self-declared white liberal spaces, as previously identified by DiAngelo (2018), that in theory aim to transform the dominant social tide by attempting to enact social justice narratives (as in the FJM), as opposed to adopting more mediocre version of social change (as in the AFM). Additionally, even when seemingly achieving social justice, including trying to institutionalize it, the 'every-day' praxis, despite intentions to the contrary, can still

Figure 7.4 Beet Chalk Board made by UCONN Art Student filled in by community members at CLiCK.

manifest oppressive and unequal power struggles. Consequently, the running of CLiCK became increasingly complex, making the questions posed in this chapter difficult to definitively answer, as will be further explored in the Conclusion in relation to the book's larger questions. Suffice it to say here that all these entanglements continued to create for me the 'non-profit conundrum', wherein a seemingly irreconcilable weave of identifiable, yet inseparable, threads created both our successes and failures, as will be discussed more in the next two chapters.

I grew up in a home where my father kept a huge vegetable garden in our backyard: lettuce, tomatoes, beans, potatoes, onions, cucumbers, garlic, peppers, cilantro, and much more.

I love to go by CLiCK to see all the planting being done in the spring and watch the progress through the summer. It brings back a lot of childhood/teen memories of Papi's garden. It brings back memories of having fresh vegetables grown in our backyard and then eating them at our kitchen table.

Middle-class Latina volunteer

Notes

1 https://www.brainyquote.com/quotes/ralph_waldo_emerson_120390
2 https://www.difficultdialogues.org/
3 CLiCK created connections with numerous small business support organizations such as CT Small Business Development Center, Windham Region Chamber of Commerce, the Northeastern CT Chamber of Commerce, and Liberty Bank in order to best support our members and to do all we could to ensure their success. Our GM worked to support member needs, while allowing them the freedom to design and implement their own businesses. However, if an individual's idea was not at all thought out, the GM would encourage them to wait and further develop the idea instead of joining CLiCK immediately.
4 https://www.smash.org/wp-content/uploads/2015/05/nonprofit_diversity_executive_summary_0.pdf
5 I came across a cooperative, Community Purchasing Alliance (CPA), that enables non-profits to band together and negotiate large utility contracts saving thousands of dollars (Prevost 2020). This approach is much like the CLiCK model in terms of members 'sharing' costs, thereby reducing them and "…allowing [nonprofits] to direct millions of dollars back towards their core mission". See https://www.cpa.coop/.
6 I thank the non-profit guru Vu Le whom I saw speak back in 2019 for using his sarcastic biting sense of humor to point out this irony. See https://www.nonprofitaf.com. Also, see numerous quotes from him later on.
7 http://www.growwindham.org/
8 http://www.growwindham.org/our-mission/
9 http://www.windhamfood.org/about/
10 See Neighborspaper.com

References

Aldarondo, E. 2013. *Advancing social justice through clinical practice*. New York: Routledge.

Bandyopadhyay, R. and Patil, V. 2017. "'The white woman's burden': The racialized, gendered politics of volunteer tourism', *Tourism Geographies* 19(4): 644–657.

Barr, M. S. 2015. 'Minority and women entrepreneurs: Building capital, networks, and skills' [online]. *The Brookings Institute, The Hamilton Project*. Available at: https://www.brookings.edu/wp-content/uploads/2016/07/minority_women_entrepreneurs_building_skills_barr.pdf [Accessed September 29, 2020]

Bejan, R. 2020. 'Robin DiAngelo's 'White Fragility' ignores the differences within whiteness' [online]. The Conversation. Available at: https://theconversation.com/robin-diangelos-whites-fragility-ignores-the-differences-within-whiteness-143728 [Accessed September 29, 2020]

Bellini, J. 2016. 'The No. 1 thing to consider before opening a restaurant' [online]. CNCB. Available at: https://www.cnbc.com/2016/01/20/heres-the-real-reason-why-most-restaurants-fail.html [Accessed September 29, 2020]

Berger, P. and Luckman, T. 1967. *The social construction of reality: A treatise in the sociology of knowledge*. New York: Anchor Books.

Berger, P. 1990. *The sacred canopy: Elements of a sociological theory of religion*. New York: Anchor Books.

Bonilla-Silva, E. 2007. *Racism without racists: Color-blind racism and the persistence of racial inequality in the United States*. Lanham, MD: Rowman and Littlefield.

Cidell, J. 2017. 'Sustainable imaginaries and the green roof on Chicago's City Hall', *Geoforum* 86(1): 169–176.

Chavis, D. M. and Lee, K. 2015. 'What is community anyway?' [online]. The *Stanford Social Innovations Review*. Available at: https://ssir.org/articles/entry/what_is_community_anyway [Accessed September 29, 2020]

Collins, P. H. and Bilge, S. 2016. *Intersectionality*. Cambridge: Polity Press.

Dempsey, M. and Brafman, O. 2018. *Radical inclusion: What the post-9/11 world should have taught us about leadership*. US: Missionday.

DiAngelo, R. 2018. *White fragility: Why it's so hard for white people to talk about racism*. New York: Beacon Press.

Duncker, D. 2020. 'The fox and the wolf: thoughts on Joe Biden's gaffe in light of Malcolm X's birthday' [online]. *The Journal Blog*. Available at: https://blog.usejournal.com/the-fox-and-the-wolf-thoughts-on-joe-bidens-gaffe-in-light-of-malcolm-x-s-birthday-827d0fcb83a4 [Accessed September 29, 2020]

DuPuis, M. E., Harrison, J. L. and Goodman, D. 2011. 'Just food?' In Alkon, A. H. and Agyeman, J., eds. *Cultivating food justice: Race, class, and sustainability*. Cambridge, MA: MIT Press, pp. 47–64.

Godfrey, P. 2015. CLiCK fundraising letter. Author's collection.

Godfrey, P. 2017. 'Reflective food truck justice: A case study in CliCk, Inc., a nonprofit, shared-use commercial kitchen'. In Agyeman, J., Mathews, C. and Sobel, H., eds. *Food truck, cultural identity and social justice: From Loncheras to Lobsta Love*. Boston, MA: MIT Press, pp. 149–167.

Ho, H. K. 2017. '8 Ways People of Color are tokenized in nonprofits' [online]. *Medium.com* Available at: https://medium.com/the-nonprofit-revolution/8-ways-people-of-color-are-tokenized-in-nonprofits-32138d0860c1 [Accessed September 29, 2020]

hooks, b. 1994. *Teaching to transgress: Education as the practice of freedom*. London: Routledge.

Lee, B. A., Martin, M. J. R. and Hall, M. 2017. 'Solamente Mexicanos? Patterns and sources of Hispanic diversity in U.S. metropolitan areas', *Social Science Research* 68(1): 117–131.

Levanon, A., England, P. and Allison, P.D. 2009. Occupational feminization and pay: Assessing causal dynamics using 1950–2000 U.S. census data [online]. SocialForces 88(2), 865–891. Available at: https://www.researchgate.net/publication/236750401_Occupational_Feminization_and_Pay_Assessing_Causal_Dynamics_Using_1950-2000_US_Census_Data [Accessed September 29, 2020]

Mangan, A. N/A. 'Earth day, white privilege and decolonizing the mind' [online]. Bioneers. Available at: https://bioneers.org/earth-day-white-privilege-decolonizing-the-mind-zmbz2004/ [Accessed September 29, 2020]

McCarthy, K. D., ed. 2001. *Women, philanthropy and civil society*. Bloomington: Indiana University Press.

McLearen, D. and Agyeman, J. 2015. *Sharing cities: A case for truly smart and sustainable cities*. Cambridge, MA: MIT Press.

McWhorter, J. 2020. 'The dehumanizing condescension of *White Fragility*: The popular book aims to combat racism but talks down to Black people' [online] *The Atlantic*. Available at: https://www.theatlantic.com/ideas/archive/2020/07/dehumanizing-condescension-white-fragility/614146/ [Accessed September 29, 2020]

Nelson, L. H. 1990. *Who knows: From Quine to a feminist empiricism*. Philadelphia, PA: Temple University Press.

Omi, M. L. and Winant, H. 1994. *Racial formation in the United States: From the 1960s to the 1990s*. New York: Routledge.

Plummer, D. L. 2017. 'Getting to we: Inclusion is more than a feeling' [online]. Huffington Post. Available at: https://www.huffpost.com/entry/getting-to-we-inclusion-is-more-than-a-feeling_b_5963ae96e4b09be68c00545b [Accessed September 29, 2020]

Provost, L. 2020. 'Connecticut churches, nonprofits band together on electricity purchases' [online]. *Energy News Network.* Available at: https://energynews.us/2020/05/28/connecticut-churches-nonprofits-band-together-on-electricity-purchases/ [Accessed September 29, 2020]

Reese, M. A. 2020. *Black food geographies: Race, self-reliance, and food access in Washington, D.C.* Chapel Hill: University of North Carolina Press.

Robinson, J. and Cole, R. J. 2015. 'Theoretical underpinnings of regenerative sustainability', *Building Research and Information* 43(2): 133–143.

Sbicca, J. and Myers, J. S. 2015. 'Food justice racial projects: Fighting racial neoliberalism from the Bay to the Big Apple', *Environmental Sociology* 3(1): 30–41.

Shapiro, J. 2019. 'The nonprofit gender gap: How a female-dominated industry still has a huge equality problem' [online]. *The Financial Diet.* Available at: https://thefinancialdiet.com/the-nonprofit-gender-gap-how-a-female-dominated-industry-still-has-a-huge-equality-problem/ [Accessed September 29, 2020]

Thurman, R. 2011. 'Non-profits don't really care about diversity' [online]. *Stanford Social Innovations Review.* Available at: https://ssir.org/articles/entry/nonprofits_dont_really_care_about_diversity#:~:text=Today's%20nonprofit%20employees%20are%20approximately, members%20are%20people%20of%20color [Accessed September 29, 2020]

Timm, J. 2016. 'The plight of the overworked nonprofit employee: Do mission-driven organizations with tight budgets have any choice but to demand long, unpaid hours of their staffs?' [online]. *The Atlantic.* Available at: https://www.theatlantic.com/business/archive/2016/08/the-plight-of-the-overworked-nonprofit-employee/497081/ [Accessed September 29, 2020]

White, M. M. 2018. *Freedom farmers: Agriculture resistance and the black freedom movement.* Chapel Hill: The University of North Carolina Press.

Williams, M. 1995. *The velveteen rabbit.* New York: Puffin Books.

Winne, M. 2013. 'The fundraising letter I'd like to receive' [online]. *Mark Winne.* Available at: https://www.markwinne.com/the-fundraising-letter-id-like-to-receive/ [Accessed July 20, 2020]

8 Thresholds of success...

Nothing is more powerful than an idea whose time has come.

Victor Hugo

The old world is dying and the new world struggles to be born: now is the time of monsters[1]

Antonio Gramsci

Stories that resonate without to within

As discussed in Chapter 4 of Part II, the question of when things begin and/or end can also complicate efforts to determine whether or not something (or someone) can be considered a 'success' or a 'failure', or whether or not something can be called 'just' or 'sustainable', as will be further discussed in the conclusion. Like the Taoist story of the old farmer who at various moments in his life is told that he has either 'such good luck' (he got a horse) or 'such bad luck' (his son fell off it and broke his leg), only to consistently respond, "Maybe", I too recognize that the potential successes or failures of CLiCK (Commercially Licensed Co-operative Kitchen) are not fixed states of being, but are, like everything, situated on a time/place-based continuum that can and will change, even from one moment, one day, one week to the next. Furthermore, evaluating CLiCK after a mere five years seems premature, not to mention highly dependent on the criteria being used and the people being asked. Nevertheless, from my positionality, I will offer what I see as some of the inseparable threads of CLiCK's successes and failures, and how in all cases we have been evolving, overlapping, and bridging between the two, conceptualized as occupying once again Keating's (2012) notion of 'threshold'. As stated in the Introduction, Keating proposes the theoretical concept of "*thresholds* [italics in original]" in order to capture "...complex interconnections among a variety of sometimes contradictory worlds – points crossed by multiple intersecting possibilities, opportunities and challenges" (p. 10). As will become increasingly apparent, any conceived immutable divisions between 'opposites' (within/without, beginning/end, success/failure, theory/practice, etc.) must in the end be recognized as illusionary, yet nevertheless still meriting ongoing reflection and evaluation in relation to larger principles. This recognition of overlap and the role of positionality also applies to the framework of just sustainabilities (JS) and its four corresponding principles, which as I previously pointed out, become increasingly complex when looked at through an intersectional lens, in that each of the principles could, and even should, be evaluated from the positionalities of all those involved. In reiterating that these are *principles*, not rigid criteria (Broto and Westman 2017), by reflecting upon CLiCK's perceived successes and/or failures from my positionality, I am going to continue proposing Keating's notion of 'thresholds', even as I seek to put forth *conceptually* tangible insights into JS in theory, in practice, and ultimately in praxis.

Perhaps one of CLiCK's greatest successes is that at this time, we are still open and still serving the diverse culinary entrepreneurial and nutritional needs of our community; after all, many predicted we would not last, due to our lack of start-up funding, the tremendous amount of work needed to make such an idea a reality, and the complex and ongoing social relationships involved in actually 'becoming Real'. Of course, we have been extremely grateful for all the public and private funding we have received, as this has been essential to our success thus far. Perhaps, more salient than the funds, however, has been the currency of the vision of CLiCK – creating a very small solution to the endemic problems of our unequal society and industrial food system and doing so *by* trying to embody and institutionalize 'radical inclusion', thereby making Real multiple aspects of the JS imaginary.

As a sociologist, I espouse a social constructionist perspective (see previous references to Berger and Luckman 1967, and Berger 1990), meaning that I consider all human experiences as socially constructed – imagined yet nevertheless real (as opposed to being 'imaginary' as discussed in Chapter 1 of Part I). Israeli historian Yuval Noah Harari, in his bestselling book *Sapiens: A Brief History of Humankind* (2015), argues that it is this ability to "…imagine things" and to do so "*collectively*" [italics in original] that has enabled our species to "rule" the world (p. 25), although of course we have not yet figured out how to rule it *equally*, nor to recognize that without all the other species who are rapidly undergoing 'the sixth extinction' (Kolbert 2014) our increasing domination will be very short lived. Furthermore, he states that humans live in "…a dual reality" – the "objective reality of rivers, trees and lions…" and "…the imagined reality of gods, nations and corporations". However, as our societies became more complex, "…the imagined reality became ever more powerful, so that today the survival of rivers, trees and lions depends on the grace of imagined entities such as United States and Google" (p. 32), which are subsequently shaped by the ways in which we 'imagine' and engage with the objective reality. Harari identifies the potential fluidity in this shaping process, writing, "Since large-scale human cooperation is based on myths, the way people cooperate can be altered by changing the myths—by telling different stories" (p. 35). Hence, human perceptions of, for example, trees can vary from one extreme to another, from being 'environmental resources' (as understood by Western patriarchal white capitalism) to 'sentient beings' (as understood by most Indigenous cultures, as well as by the eclectic perspectives of ecofeminists – see Diamond and

Orenstein 1990) that can communicate, as intriguingly argued more recently from a Western scientific perspective (as opposed to a strictly spiritual and/or philosophical one as in the past) by the forester Peter Wohlleben (2016). These variations are of course dependent on our larger culturally embedded imagined realities, which shape and limit our abilities to perceive the natural world, as well as our actual day-to-day behaviors, including whether we deforest the planet for profit or recognize its intrinsic worth spiritually, emotionally, and of course physically.

The idea of 'telling different stories' ties directly to another aspect of CLiCK's success: right from the beginning, Tina and I made a conscious decision to tell and grow a different kind of story, a story of grassroots *solutions*. Long before CLiCK existed as an actual entity, we 'self-deceiving' idealists told ourselves and others what we thought it was going to be like and how 'good' it was going to be. This intentional optimism, verging on 'magical thinking' (that can overlap/ be perceived as self-deceiving), which we extended to Facebook, Instagram, and eventually our professionally polished website (a professional website which somewhat belies our grassroots reality), was in opposition to the more prevalent story on the political left, which tends to focus on the *problems* that seemingly can only be addressed by long-term, full-scale solutions. Calls for 'capitalism to end' thereby become dead ends in and of themselves given the scale of the demand, as opposed to taking small steps toward 'new models' that, as mentioned, will '...make the existing model obsolete'. We felt that if we stayed exclusively focused on fighting the problems, we would not be in a position to create what is new, starting with a 'new story'.

However, 'telling different stories', in particular stories that attempt to spin visions of a more just and sustainable world, is excessively challenging, especially when many politicians deny social and environmental problems outright (Bowers 1997) and political discourse predominantly consists of oppositional politics. As Keating (2012), argues, "...oppositional politics and oppositional thinking have not enabled us to radically transform society ...we remain immersed in conflict— conflicts which subtly reinforce the very systems against which we struggle" (p. 3). In contrast, as mentioned, she proposes, "'threshold theories', to underscore their nonbinary, liminal, potentially transformative status" (p. 10). This framework is meant to "...inspire us to be bold, to dream big, to affirm the possibility of transformation, to envision radical change", as threshold theories "...mark crisis points, spaces where conflicting values, ideas, and beliefs converge..." (p. 11). As such, it might be appropriate to propose that 'thresholds' are both the time of magic, and as Gramsci alluded, 'the time of monsters', wherein 'the old world is dying and a new one is struggling to be born'. Additionally, Keating's (2012) concept of 'threshold theories' is fitting for our attempts at making CLiCK a 'new collaborative model', one that is nevertheless still part of the old one in complex and contradictory ways, and yet is based on an attempt to collectively 'envision radical, relational, change'. This is of course as long as we stay vigilant, for the 'old ways' are dominant, as already previously proposed, persistent, and ever ready to reinsert/reassert themselves through us and around us.

As stated in the Introduction, CLiCK is by no means *alone* in attempting to 'birth a new model' by collectively 'envisioning radical, relational, change' in these 'monstrous and magical' thresholds between worlds. Most organizations that are part of the Food Justice Movement (FJM) would also fit these descriptors, as well as many that are part of what Paul Hawkins (2007) has referred to as the largest social movement in the world, consisting of "…over one million—and maybe even two—million organizations working toward ecological sustainability and social justice" (p. 2). For Hawkins, all these global social justice organizations are symbolically acting as humanity's "…immune response to toxins like political corruption, economic disease, and ecological degradation" (p. 142). Moreover, even before the global outbreak of COVID-19 and the recent reigniting of the #BlackLivesMatterMovement, increasing numbers of people around the world have understood that 'the old world is dying'. Of course, when Gramsci wrote his words, he was in prison under the fascist regime of Mussolini, and although one could think that his notion then of 'the old world' did eventually die, I would argue that culturally it is the same 'old world' that continues to die. For now, not only does the 'old world' still include "…a globalized industrial economy based on ceaseless resource extraction" but also now more than ever we are witnessing the collapsing of "…the ecosystems on which we depend" (Danaher et al. 2007, p. 2).

Kevin Danaher, Shannon Briggs, and Jason Mark, in their book *Building the Green Economy: Success Stories from the Grassroots* (2007), argue, "If we want to avoid ecological disaster—and the social catastrophes that will come [are already here] with it—we must create a way of living that is more deeply connected to nature" (p. 2). Like Hawkins (2007), their focus, as their title suggests, is on stories of "…pioneers of this local, green economy movement…" (p. 2). These people, like us at CLiCK, "…aren't pie-in-the-sky prophets. They are hard at work, on the ground", engaged in a variety of regenerative activities including "grow[ing] organic food" (p. 2). Additionally, they/we recognize that all of these initiatives are ultimately "…about knitting together communities…" and doing so from within one's own communities, while engaging both "…ecological thinking…" (p. 6) that recognizes the interconnectedness of all life, as well as the power of "…imagination". The authors seek to be "…movement storytellers…" and propose that the stories they tell illustrate "…what we like to call 'the alchemy of empowerment'" (p. 8), wherein "concerns" become "commitments", "passive consumes" become "active citizens" and a "magic spell" makes people "see themselves as heroes of the story" (p. 8). Crucially, however, they reveal that this 'spell' is of course "…not all that magical…" and that the "…philosopher's stone, we guess, is you…" as it is through "…coming together [that] people find a way to tap into that most important of our renewable resources—human creativity" (p. 9).

I have quoted extensively from Danaher et al. (2007) primarily because their research affirms my own and places my story/CLiCK's story in a larger social context as part of a national–global grassroots movement of people hard at work trying to figure things out and making things up as they go along. Thereby, we are making our stories simultaneously more mundane but also more 'magical' (despite their caveat, as will be further explored). Furthermore, they explicitly state that,

> ...we've had enough of the bogeyman fright tales, the reports of how bad everything is ...let's stop telling *their* story [italics in original]—the stories of abuse and corruption and injustice—and start writing our own...Enough of hand-wringing. It's time for fists in the air.
>
> (p. 7)

Of course, as discussed, we urgently need these new models. My only critique, however, would be that their argument lacks an explicit and centralized analysis of racism, sexism, economic inequalities, or their intersections, although they do include an interview with environmental justice activist Gopal Dayaneni, who mentions these issues (partially doing their work for them). Therefore, although the themes of 'the alchemy of empowerment' and 'fists in the air' are one that I recognize and support, I would argue that a more radical and resonant way of framing the necessary struggles would be to explicitly name the so-called 'bogeyman' before moving onto the 'successes'. Danaher et al. (2007) do recognize that "Humans define themselves with stories...", another key reason why I draw upon their work; they note that "...shared stories form the backbone of tradition and culture" (p. 7), as do conversations, which they describe as "...the wellspring of community and a fundamental ingredient of democracy" (p. 13). This of course links back to Berger and Luckman (1990), as well as Harari's proposal that it is our collective storytelling abilities that made "Sapiens rule the world, whereas ants eat our leftovers and chimps are locked up in zoos and research laboratories" (p. 25). Of course, from an intersectional perspective, we need to highlight *which* sapiens rule the world and critique the manner in which they do so in relation to all the others, including all the other species as mentioned, but his point is nevertheless still valid. In a much *much* smaller way, storytelling has helped CLiCK achieve success, but in a cooperative and collaborative and mutually beneficial way, as opposed to the dominating manner exposed by Harari. Hence, as King (2008) was quoted earlier warning, we must "...be careful with the stories [we] tell", and the ones we "are told" (p. 10).

To offer a more specific example, we learned early on that crafting successful grant applications requires following the words of the cheerfully sarcastic nonprofit consultant Vu Le: "...tell funders exactly what we think y'all want to hear, sugarcoating everything in jargon and BS" (Le 2018). This does not mean lying; rather it means recognizing that the 'sugarcoating' and 'BS' are forms of telling a *story* that speaks to the *collective imaginary* of what the funders/society in general believe to be important and therefore what *should* already exist but does not. In short, successful grant writing, just like successful community building, requires *collectively* enacting, in the words of French writer Victor Hugo, "...an idea whose time has come", an idea that is already socially true and therefore collectively real in all but physical reality.

Figure 8.1 Sananni Bishwakarma doing a Nepali Cooking Class.

*Their mere survival is a testimony to dedication by founders, current managers
as well as a community of supporters.*
  Middle-class white male, former Co-op board member

Stories that resonate within to without

Instrumental in helping me in my role as board president and in writing grants
'telling funders exactly what we think y'all want to hear' was our second GM,
whom we hired after our second year. As a trained chef, culinary instructor,
and emerging entrepreneur, she had first heard of us when seeking to start her
own business with locally sourced (to the extent possible, given seasonal limits)
chef-prepared meals-to-go. Given her culinary background (as opposed to the nutritional background of our first GM), she brought a number of new initiatives,
such as further developing our culinary education programming, beginning to
formalize our policies and procedures (kitchen/members handbook, use prices,
health and safety, best practices, etc.) and doing extensive outreach to small local
farmers to publicize the opportunity to create value-added products and reduce
food waste, in line with the original goals from the Willimantic Co-op. Being
biracial, with family members from Puerto Rico (PR), as well as White family

members from Connecticut, she presented CLiCK with a closer connection to the local PR community. However, in my opinion, this was more of a symbolic connection given that she did not overtly identify as Latina, was not bilingual, and grew up middle-class in a predominantly white rural CT town (not Willimantic) and therefore did not personally share the economic/cultural struggles, hence identity markers, of local lower-income Puerto Ricans. Nevertheless, even this tenuous more culturally inclusive link to the surrounding Latinx community brought us more in line with our social justice values for staff diversity, as well as her own alignment with CLiCK's mission and our social justice goals.

As an emerging culinary entrepreneur herself, our new GM understood and could empathize with the struggles of our members trying to navigate the food regulatory system, creating a relationship of empathy and solidarity. At the same time, this 'solidarity' blurred the lines between her roles as CLiCK's GM and as her own entrepreneur, an entanglement that the board could never quite unravel. The balance ultimately tipped in favor of her own business, which continues to operate out of CLiCK's kitchens, enabling her to step down from being CLiCK's GM to focus full-time on her business. However, during the years when she was CLiCK's GM and I was the board president, we spent many hours in formal and informal meetings, sharing what Meredith E Abarca (2004) refers to as "*charlas culinarias* (culinary chats)" (Abarca 2004, p. 1; also in Godfrey and Torres 2020, p. 281). These chats, although conducted in English, had an intimacy about them that I find is captured perfectly in Abarca's Spanish term; they focused on our joint overseeing the daily operations of CLiCK and included problem solving (which on any given day might involve addressing issues with members, going over plans to build an accessible toilet for the sake of the public and a board member who uses a wheelchair, a lack of funds for cleaning products, etc.), as well as event planning, fundraising, education outreach, and grant writing. Yet, what stands out most in my memory was sharing our beliefs in relation to 'justice' and 'sustainability', imbuing our decision making with these beliefs, and thereby sharing our tangible actions to further make physically real, hence institutionalize, the 'JS imaginary'.

I believe these weekly 'chats' helped us both to grow in our roles and subsequently helped CLiCK to grow, leading to an increase in our number of kitchen members. Over the years, our members have come from over ten different countries such as Mexico, Thailand, Sudan, South Korea, and Puerto Rico, among others. With a range of educational backgrounds, culinary skill levels, and starting capital, members all have been given opportunities to share CLiCK's kitchens and to turn their recipes into food commodities for sale. Business ideas and start-ups (including a number of food trucks) which have used CLiCK include vegan meals, gluten-free baking, dried soups, granola bars, hot sauces, kombucha, gourmet chocolates, jams and jellies, and meals-to-go, as well as those businesses that have focused on national/cultural foods.

With all this successful growth, it was apparent we needed to hire another part-time person to help with the finances, to act as an administrative assistant (AA) and to help oversee the teaching kitchen (we still have our former vice president [VP] offering classes part time). Upon the recommendation of the GM, as well as a result of the board conducting numerous interviews, we ended up hiring a white woman who was also incubating a business out of CLiCK. Bringing a variety of talents, including an extensive knowledge of heritage foods (her food business focused on Renaissance Fairs, while adapting recipes to be modern as in for example gluten-free), our second employee, albeit also part time, made us on the board feel that we were making progress and could step back a bit more from being a 'working board' to being a 'governing one'. However, given that both were part time and both were also using the CLiCK kitchens, our sense of euphoria was short lived as now we entered a new phase of employee conflicts (not just board and organizational ones), member favoritism (some members liked the GM and not the AA, others the other way round), kitchen use discrepancies, and many other unexplainable happenings, interspersed with brief moments of collective inspiration and synergy. As a result, it further divided members of the board as each employee had an assigned board member (the GM our VP and the AA our treasurer) to oversee them and yet the phrase that became our mantra was that without an executive director (ED) who could oversee more regularly both the GM and AA there would be 'no one minding the store'; hence, the indecipherable threads of unknowns would continue to be woven. As board president, I found myself trapped for almost two years within what became an increasingly messy web made up of multiple realties, truths, and tales. Additionally, seeking a more inclusive cooperative values-based engaged leadership style, that as mentioned, one board member frequently described as 'loosey goosey', I encouraged the board not to intervene or assert our authority over them but to try and allow things to work themselves out. Unfortunately, things did not work themselves out, resulting in the AA quitting of her own accord, further adding to the board's sense that we never knew *whose* 'truth' was 'true'. And yet although we felt a sense of personal failure in not having successfully resolved our employee conflicts, once again we were not alone but rather part of a global trend. According to a CPP Global Human Capital Report (CPP 2008) "…the majority of employees (85%) have to deal with conflict to some degree and 29% do so 'always' or 'frequently'" (p. 3). As for the causes, "…workplace conflicts are seen as personality clashes and warring egos (49%), followed by stress (34%) and heavy workloads (33%)" all of which were highly applicable in our case. We had two highly motivated female entrepreneurs working on their own food businesses and both working part time for CLiCK in what were actually full-time positions for inadequate pay, while simultaneously attending to the diverse needs of multiple other entrepreneurs who were at various stages of business incubation. The results were not surprisingly highly unpredictable, combustible, and nevertheless uniquely energizing, bringing to CLiCK aspects of what felt like a dysfunctional, yet intimately woven family.

As mentioned in the previous chapter, there is tremendous flux in the food business industry and so ideas, proposals, and people have come and gone. We might get excited about three new members joining CLiCK, only to find that two would slowly disappear and the third would start and then only use a few hours a week. On a number of occasions, such members would be seen at the farmers market selling a product and claiming it was made at CLiCK, when they had not booked any recent kitchen time. It soon became apparent that several were using CLiCK in name only – joining and then telling the North Central District Health Department that they were working out of CLiCK in order to get their food vending license – but continuing to cook from home or elsewhere. Due to the potential for public health impacts, we informed the health department when these incidents occurred, but in general had limited ability to redress them. We did put up notices appealing to our cooperative values and their roles as members, thereby hoping to make them more accountable to the collective and less likely to 'cheat'. It was never clear if we were successful, as of course the financial pressure on start-up businesses, which are endemic to the corporate economic model, can be said to incentivize cheating and certainly contributed to our small-scale experiences of Garrett Hardin's concept of the 'tragedy of the commons' (1968). However, as Hedley Freake (long time board member) and I (Godfrey and Freake 2016) have previously noted, in line with Nobel Prize winning economist Elinor Ostrom's challenge to Hardin's overgeneralized 'tragedy', CLiCK has for the most part experienced from its members a cooperative and mutually beneficial use of its common resources.

Then, in 2018, Connecticut passed the Cottage Law[2] allowing for small-scale home/farm production of certain items (such as baked goods, jams & jellies, and cereals), which resulted in the loss of a number of CLiCK members who now had no need for a commercial kitchen. We supported this law, but it did create additional competition for CLiCK as a facility, although the law was limited to items directly sold to customers as opposed to a third-party vendor. Our GM was very familiar with these state and federal rules and regulations and was a key asset both in helping members navigate them as well as imagining new ways for CLiCK to grow within the regulatory landscape. One ongoing goal is for CLiCK to become an official Co-Packer (able to process food items for farmers rather than just providing the space for them to do it themselves) and even one day a United States Department of Agriculture (USDA)-certified kitchen (which would require significant structural changes, including providing the USDA inspectors with an office space and designated toilet when they come to inspect!). I will say more about these ideas in the section on 'Failures', but they are mentioned here as they were all explored during conversations about how to best further ensure CLiCK's success and serve our local farmers.

To pursue the Co-op's original vision to support local farmers by providing them with a licensed kitchen in which to make value-added products, we secured a number of Connecticut Department of Agriculture grants. Connecticut has seen a 60% increase in the number of farms since the 1980s and currently has about 6,000 small farms; the average farm size is about 73 acres (Keough 2017), although among the farmers we know the average size is much smaller. One of the Connecticut Department of Agriculture grants funded our proposal to pilot a program for processing locally grown produce for the public schools in Windham. CLiCK, along with many others in the food movements (both the AFM and the FJM), as well as the CT Department of Agriculture, have all looked to 'locally grown food' (defined by the USDA as within 400 miles)[3] as the panacea for the ills of the industrial food system. And in many ways, food that is produced and sold locally is done so by local businesses (farmers, chefs, restaurants…etc.), and a comprehensive review of research studies all comparing differences between big box national stores and small locally owned ones has shown that buying from the small local one has many benefits (Michell 2016). For example, it helps support start-up businesses (such as those at CLiCK), reduces inequality/improves wages, keeps a larger percentage of funds in local circulation – including taxes, helps create jobs, and enhances social capital and well-being (Michell 2016). However, as with all simplistic solutions, in particular when it comes to the complexities of food production, calling for merely 'local solutions' is grossly inadequate.

Branden Born and Mark Purcell (2006) have proposed the term "…the local trap…" to refer to the assumption that "…local-scale food will be inherently more socially just than a national-scale or global-scale system" (p. 195). An extreme example of the 'local trap' that they identify are the "…buy local campaigns… since they uncritically conflate so much with localization" (p. 200) – promising everything from sustainability, to social justice, to better nutrition, and food security – much as we have done with our storytelling about CLiCK. Born and Purcell continue, "No matter what its scale, the outcomes produced by a food system are contextual; they depend on the actors and agendas that are empowered by the particular social relations in a given food system" (p. 196), while also seemingly embodying a nostalgia for being connected to the land where we live.

However, despite all these assumed positive outcomes, including nostalgia for being connected to the land where we live, there is nothing inherently *just* about 'the local'. In fact, logically, it can be the site of as much social inequality as anywhere else (and of course everywhere is someone's 'local'). Therefore, seeking to find a balance, E. Melanie DuPuis and David Goldman (2005) argue that what is needed is not "…reductionist global-local binaries" in which the global is considered "…the domain of capital while paradoxically framing the local as a site of empowerment" (p. 369), but rather "…an inclusive and reflexive politics of place…" that is "…a mutually constitutive, imperfect, political process in which the local and the global make each other on an everyday basis" (p. 369). This description is much more in line with the reality of CLiCK's *practice*, despite us also intentionally perpetuating the 'local food' narrative everyone wants to hear.

When we first opened, we had many discussions at board meetings about how we could encourage members to support our cooperative and social justice values in their business practices, especially business practices related to from where members purchased ingredients and the resulting foods members produced. One of our first members, a company called *Nourish*, did try to incorporate social and environmental ethics into their business goals, but despite making vegan gluten-free noodles (not local) with locally sourced organic vegetables, organic black bean (not local) brownies, organic avocado (not local) ice-cream, in the end they could not financially sustain themselves for more than a year as they purchased and produced in such small quantities. As other members joined who did not seek to embody these visions of using only organic, and/or local, and/or ethically sourced ingredients, we accepted that our ideals were not economically viable, nor geographically achievable, nor socially inclusive, and as such they could not be 'just'.

To give an example, perhaps one of CLiCK's best success stories is that of our kitchen member Maria, who is an immigrant from Mexico (Figure 8.2). As detailed in my co-authored chapter (Godfrey and Torres 2020), before coming to CLiCK, Maria would cook hundreds of meals a day in her own home and drive them out to Mexican farm workers for sale. This 'underground' operation was not only illegal but also meant that

> ...her living room had vats of rice, beans, and tortillas in places where most people have furniture. The walls of her apartment were covered in steam and grease and her children were teased at school because they 'always smelled like food'.
>
> (p. 287)

What took her "...10 hours to cook in her house now takes only 5 hours..." (p. 287) at CLiCK, and she has more employees and the support of CLiCK staff and other members. As for her ingredients, although Maria herself now lives locally, most of

Figure 8.2 Maria de los Angeles Cooking.

what she uses in her cooking, along with her recipes, come from Mexico. When my co-author (who was also translating) and I were interviewing her, Maria explained, "You know there are food smugglers …They cross the border and bring back items we can't get here… [these items] …make my food authentic" (p. 286). As such, in providing Maria with an affordable kitchen, financial and business support, and a radically inclusive community, 'a mutually constitutive, imperfect, political process in which the local and the global [do] make each other on an everyday basis' is being served at CLiCK – as is some amazingly delicious Mexican food.

Of course, bringing foods from Mexico may be considered unsustainable in terms of 'food miles' (Lang and Heasman 2004), but as Pierre Desrochers and Hiroko Shimizu argue in *Yes, We Have No Bananas: A Critique of the Food Miles Perspective* (2008), this view "…ignores productivity differentials between geographic locations…" and "…assumes that producing a given food item requires the same amount of inputs independently of where it is produced" (p. 6). This assumption obviously does not hold up to scrutiny; in Mexico, for example, there are two growing seasons versus one in Connecticut, and so the growing of habanero peppers in Mexico is a lower-input production than year-round production in Connecticut would be. Hence, in a parallel to the idea of the 'local trap' in terms of popular favorable prejudice, Desrochers and Shimizu (2008) conclude, "…the benefits [environmental, economic, and social] claimed by food-miles proponents have little basis in fact" (p. 64). A key example is referenced in the title of their policy brief: while "…it is possible to grow bananas in Iceland, this was never done on a large scale because they cost much less when shipped in from tropical countries" (p. 4). This issue of cost when trying to purchase and process/add value to local food, despite our goals for our CT Department of Agriculture grants, was one that continually impeded CLiCK's ability to sustain such practices after the grant funds ran out, as will be explored in the next chapter on failures.

In trying to meet the goals of the CT Department of Agriculture grants we were awarded, I would say that we fell into the 'local trap', believing, as Freake and I wrote back in 2016, that CLiCK would create "…pathways between the local production of raw fruits and vegetables by small scale farmers and the marginal market demand for locally processed food products" (p. 125). We were encouraged by reports of other examples, such as what we had seen at the Franklin County Community Development Corp (CDC) previously mentioned, where the center has the capital to purchase in volume from farmers for processing and the equipment to reduce the reliance on labor. Additionally, we looked to reports of "community supported canning" wherein farmers would come together to engage in small batch processing to supplement their Community Supported Agriculture (CSA) offerings (Rudalevige 2013). However, unlike these two different examples, we neither had the capital to purchase the required volume of produce, nor the level of equipment required to reduce the use of labor, nor were the farmers in our area able to do the processing themselves and therefore looked to us to do it for them. Hence, we encountered an additional and more literal version of the 'local trap', discovering after a year or so that small-scale processing of 'local food' for value-added sale is a '*labor* trap' that a small capital-poor kitchen like CLiCK could not escape. In fact, it seems that the only way to fully escape it within the current capitalist system is by scaling up in order to keep the price of food down, as Holt-Gimenez (2017) argues, through "mechanization" and labor that is "…super-exploited, being paid wages too low to support themselves and their families at an average standard of living…", and including "…undocumented labor" (p. 64). As I will discuss in more detail in the next section, the *success* here has been our realization that our struggles in trying to make processing locally grown foods a viable option, even on a small scale, are not of our own making. For as Freake and I realized, "…we are literally swimming against the corporate tide" (Godfrey and Freake 2016, p. 125), a tide in which, as Holt-Gimenez identifies (2017), industrial agriculture dominates and "favors large farms—organic or otherwise" (p. 67). This is not to claim that sustaining and growing CLiCK as a co-packing/processing facility can never be done, or that other kitchens have not successfully achieved this goal in ways that are not as problematic as Holt-Gimenez describes, but that as of yet this part of CLiCK's story has not yet 'become Real'. In fact, as Freake and I stated, our problem is *not* "…our model of seeking to support small scale farmers through local processing, small food business creation and health and nutrition education …" (Godfrey and Freake 2016, p. 126). Rather, as we go on to say,

> The problem is the larger industrial food system that like Goliath has a great advantage over David, especially a very small and recently established David, and therefore makes creating even a small change very challenging. Regardless, what CLiCK has on its side, like the original David, is righteousness in terms of not only speaking about a more just, equitable, sustainable, held in common, food system but more audaciously of trying to create one.
>
> (p. 126)

Such 'righteousness' makes an excellent story, and so, let that in and of itself be seen as a 'success', both symbolic and otherwise.

Another fluctuating success at CLiCK has been our nutrition programming, which is intended to represent 'just nutrition', in that "…issues of social class, racism and gender are understood as important elements that influence food access and nutritional outcomes" (Godfrey and Freake 2016, pp. 123–124). These intersecting identities are, as we go on to say, "…important predictors of health outcomes, such as diabetes, cardiovascular disease and cancer via pathways that are both nutrition-dependent (overconsumption of cheap and filling foods of limited nutritional value) and independent (stress, access to health care)" (p. 124). As with the idea of encouraging our members to 'buy local', we also hoped to encourage them "…to consider the nutritional benefits associated with their products along with the sourcing of their ingredients", but this also did not pan out for similar reasons. What we have attempted instead has been an institutional emphasis on 'healthy eating' that is also culturally appropriate and open to both ideas emerging from CLiCK's own educational programming and from the initiatives of community members. For example, a vegan couple, both of whom are physicians from Mexico who work at the local community medical center, offered an eight-week course in Spanish on vegan cooking to help their patients address issues of obesity and diabetes, which are higher among Hispanics than for non-Hispanic whites (HHS.gov 2020; HHS.gov 2019). Additionally, programs like Cooking Matters,[4] Weight Watchers,[5] and University of Connecticut (UCONN) Extension's Expanded Food and Nutrition Education Program,[6] among others, have used our teaching kitchen and educational space to offer their own versions of healthy eating/living. Although they may not all entirely align with all aspects of CLiCK's social justice vision, they have all added to CLiCK's role in the community and to a culture of inclusion. In this manner, "CLiCK is not only part of a physical commons providing shared commercial kitchen space but also of the knowledge commons, creating a repository of shared food practices that can benefit the whole community" (Godfrey and Freake 2016, p. 124).

A final noteworthy success, which may or may not be sustained over the next year, has been CLiCK's hiring of an executive director (ED). In my time as the board president, I had essentially played the role of ED, and so it had been an ongoing goal for us to find the funds to support an ED staff position so that I could step down and focus on writing this book. Early in 2019, I realized that, ED or no ED, I could no longer fulfill my duties as board president and had to find a way to exit. I also knew that it was time to leave to avoid what Block (2016) has referred to as "founder's syndrome" (p. 135), in that those who found organizations, particularly non-profits given their emphasis on a larger social vision, can become impediments to the organization's future growth. And so it happened that one morning during this time, I woke up to clearly announce to Tina that a friend of ours (a white female) – who had actually been an early CLiCK board member when we first started and who had recently left her position as ED of the local No Freeze Shelter, a homeless shelter run during the winter where she focused on finding housing, building community, and practicing radical inclusion – should become CLiCK's ED, and that the Jeffrey P. Ossen Foundation, who had supported us extensively in the past, would pay for it. The idea, and its clarity upon waking, felt so right to me (especially because my friend was well known to and admired by the foundation's Director) that I sent the foundation's Director an email excitedly proposing it.

Of course, she turned the idea down, since the Foundation had already given us ample funds and asking for funds in this manner was wildly out of the ordinary. Nevertheless, tapping into my tendency for magical thinking, I was still confident that it would work out. Of course, my friend still had to agree (which she did), as did the board (which they did), and ultimately the funder had to agree, but all of these hurdles seemed small in comparison to the power of my convictions. Then, the foundation Director went to Spain to walk the Camino (notorious for prompting soul searching), and I said to Tina, "When she comes back, she will call me and say that she has changed her mind". I am happy and extremely grateful to say that in fact when she returned, she called me and ended up, with her board approval, offering to pay for six months and later renewing the

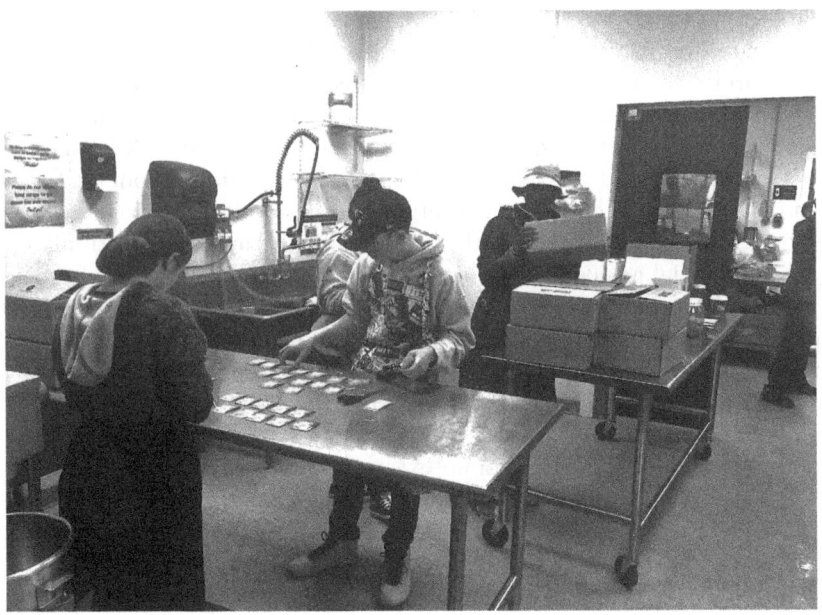

Figure 8.3 Youth from Local Food Organization Working in Kitchen.

position for another six months, which is where we are now. I am hopeful that either she will find it in her heart and in her foundation to do so again or that some other funds will appear (as they always seem to have done), as the ED has made great strides in helping to continually grow CLiCK's JS imaginary, as from her years of successfully working for and *with* the homeless population and prioritizing housing, she knows how to further create a culture of radical inclusion. How this will play out remains to be seen and will of course further change how CLiCK evolves and changes and therefore how it is experienced, understood, and evaluated.

> *I think it needs to move from a place that people have heard of and think is a good idea to one that they visit regularly and take ownership. We should offer community memberships with clearly defined benefits, including attractive programming that shares knowledge and builds community. We need to strengthen the relationships between CLiCK and other local organizations. This has been attempted, but with only partial success. All this depends on its economic viability, so we do need to attract more commercial kitchen members and help them be successful.*
> Middle-class white male board member
>
> *Working on bringing back the connections to the community CLiCK had. Reaching out to people to remind them that CLiCK is part of them and for them.*
> Middle-class white female employee

Notes

1 This is a popular loose English translation – the literal one is "*The crisis consists precisely in the fact that the old is dying and the new cannot be born; in this interregnum, a great variety of morbid symptoms appear*" from the Italian "*La crisi consiste appunto nel fatto che il vecchio muore e il nuovo non può nascere: in questo interregno si verificano i fenomeni morbosi più svariati*".
2 https://www.cga.ct.gov/2018/rpt/pdf/2018-R-0190.pdf
3 https://www.ers.usda.gov/amber-waves/2010/december/varied-interests-drive-growing-popularity-of-local-foods/#:~:text=According%20to%20the%20definition%20adopted,in%20which%20it%20is%20produced.
4 See https://cookingmatters.org/
5 See https://www.weightwatchers.com/us/
6 See https://efnep.uconn.edu/

References

Abarca, M. E. 2004. 'Authentic or not, it's original', *Food and Foodways* 12(1): 1–25.

Berger, P. and Luckman, T. 1967. *The social construction of reality: A treatise in the sociology of knowledge.* New York: Anchor Books.

Berger, P. 1990. *The sacred canopy: Elements of a sociological theory of religion.* New York: Anchor Books.

Block, S. R. 2016. 'Founder's syndrome in nonprofit organizations'. In Farazmand A., ed. *Global encyclopedia of public administration, public policy, and governance.* New York: Springer.

Born, B. and Purcell, M. 2006. 'Avoiding the local trap: Scale and food systems in planning research', *Journal of Planning Education and Research* 26(1): 195–207.

Bowers, C. A. 1997. *The Culture of Denial: Why the Environmental Movement Needs a Strategy for Reforming Universities and Public Schools.* SUNY Press.

Broto, V. C. and Westman, L. 2017. 'Just sustainabilities and local action: Evidence from 400 flagship initiatives', *Local Environment* 22(5): 635–650.

CPP Global Human Capital Report, 2008. *Workplace conflict and how businesses can harness it to thrive.* [online] Available at: https://www.themyersbriggs.com/download/item/f39a8b7fb4fe4daface552d9f485c825 [Accessed September 29, 2020]

Danaher, K., Briggs, S. and Mark, J. 2007. *Building the green economy: Success stories from the grassroots.* Sausalito, CA: PoliPointPress.

Desrochers, P. and Shimizu, H. 2008. 'Yes, we have no bananas: A critique of the food miles perspective' [online]. *Mercatus Institute.* Available at: https://mercatus.org/uploadedFiles/Mercatus/Publications/Yes%20We%20Have%20No%20Bananas_%20A%20Critique%20of%20the%20Food%20Mile%20Perspective.pdf [Accessed September 29, 2020]

Diamond, I. and Orenstein, G. F., eds. 1990. *Reweaving the world: The emergence of ecofeminism.* San Francisco, CA: Sierra Club.

DuPuis, M. E. and Goldman, D. 2005. 'Should we go "home " to eat?: Toward a reflexive politics of localism', *Journal of Rural Studies*, 2: 359–371.

Godfrey, P. and Freake, H. 2016. 'Feeding community: A case study of a shared-use commercial kitchen in eastern Connecticut'. In Bosso, C., ed. *Feeding cities: Improving local food access, sustainability, and resilience.* London: Routledge, pp. 113–128.

Godfrey, P. and Torres, D. 2020. 'Recipes for immigrant lives: Crossing, cultivating, cooking, and culture at a shared-use commercial kitchen'. In Giacalone, S. and Agyeman, J., eds. *The immigrant-food nexus: Borders, labor, and identity in North America.* Boston, MA: MIT Press, pp. 281–296.

HHS.gov. 2019. U.S department of health and human services office of minority health. Diabetes and Hispanic Americans. Available at: https://minorityhealth.hhs.gov/omh/browse.aspx?lvl=4&lvlid=63 [Accessed 29 September 2020].

HHS.gov. 2020. U.S department of health and human services office of minority health. Obesity and Hispanic Americans. Available at: https://minorityhealth.hhs.gov/omh/browse.aspx?lvl=4&lvlid=70 [Accessed 29 September 2020].

Harari, Y. N. 2015. *Sapiens: A brief history of humankind.* New York: Harper.

Hardin, G. 1968. 'The tragedy of the commons', *Science* 162(1): 1243–1248.

Hawkins, P. 2008. Blessed unrest: How the largest social movement in history is restoring grace, justice, and beauty to the world. New York: Penguin Books.

Holt-Gimenez, E. 2017. The foodies guide to capitalism: Understanding the political economy of what we eat. New York: Monthly Review Press.

Keating, A. L. 2012. *Transformation now!: Towards a post-oppositional politics of change.* Ann Arbor, MI: University of Illinois Press.

Keough, G. 2017. Farming on the rise in Connecticut. New England state statistician, national agricultural statistics service in conservation. Available at: https://www.usda.gov/media/blog/2014/06/12/farming-rise-connecticut [Accessed 29 September 2020].

King, T. 2008. *The truth about stories: A native narrative.* Minneapolis: University of Minnesota Press.

Kolbert, E. 2014. *The sixth extinction: An unnatural history.* New York: Henry Holt and Co.

Lang, T. and Heasman, M. 2004. *Food wars: The global battle for mouths, minds and markets.* London: Earthscan.

Le, V. 2018. *Answers on grant proposals if nonprofits were brutally honest with funders.* [online]. Available at: https://nonprofitaf.com/2018/02/answers-on-grant-proposals-if-nonprofits-were-brutally-honest-with-funders/ [Accessed 29 September 2020].

Michell, S. 2016. 'Key studies: Why local matters'. [online]. Available at: https://ilsr.org/key-studies-why-local-matters/ [Accessed 29 September 2020].

Rudalevige, C. B. 2013. 'Community supported canning gets locavores through winter'. [online]. *The salt-What's on your plate.* Available at https://www.npr.org/sections/thesalt/2013/10/29/241664654/community-supported-canning-gets-locavores-through-winter [Accessed September 29, 2020]

Wohlleben, P. 2016. *The hidden lives of trees: What they feel, how they communicate.* New York: Greystone Books.

9 Failures and unknowns (thus far...)

Light will someday split you open; even if your life is now a cage.

Hafiz

Old models, monsters, and ideas whose time has not *yet* come from within to without

One abiding theme in telling my story of CLiCK (Commercially Licensed Co-operative Kitchen) has been the role of time and the questions as to how and when we can evaluate the measure of any initiative and in particular those in relation to just sustainabilities (JS) and its four principles. For, as the poet Hafiz eloquently claims "...even if your life is now a cage", someday "Light will ... split you open",[1] thereby totally transforming it. Thus, CLiCK's failures can be said to be the result of a combination of the ubiquity and persistence of our 'cages', our 'old models', our 'monsters', and the challenges of birthing ideas 'whose time has *yet* to come'. Of course, there were identifiable individual failures (my own, as well as those of others) as a result of various intersections of ignorance (...), mundane stupidity, arrogance, and prejudices, but the advantage (and disadvantages) of working collectively is that such individual failings get absorbed into the collective and no one individual is ever held up for blame. That said, as the co-founder and board president, I often felt personally responsible for CLiCK's success and/or impending failure, hence my past experiences of intense anxiety, while at times, I wanted to hold other individuals accountable for failings that have compromised aspects of our mission and/or immediate goals, including our finances. Nevertheless, I constantly reminded myself of the Latin saying "Unus pro omnibus, omnes pro uno", made famous by Alexandre Dumas in his bestselling novel *The Three Musketeers* published in 1844 (1997)– "All for one and one for all". As a collective experiment, where we have tried to embody the adage I learned from The People's Supper that, "Social change moves at the speed of relationships, relationships move at the speed of trust" (Bailey and Flowers 2016), we have had the collective freedom to try ideas, to push the boundaries of what we believed was the JS imaginary, and to experience small failures, in both the short term and over time, some of which may nevertheless become future successes. Additionally, in positing, as I have previously, that CLiCK represents an "imperfect" example of "reflexive food justice" (DuPuis et al. 2011, p. 297; Godfrey 2017, p. 152), it is vital that I offer up examples of our failures – our monsters and/or our ideas whose time has not *yet* come.

One of CLiCK's initial successes that I now see as one of CLiCK's emerging failures has been the loss of the energized, highly intentional, radically inclusive community outreach events, which (as discussed in the last chapter) constituted our earlier attempts to make CLiCK Real, as in beloved by and of our diverse local communities. These were, as mentioned, very time-consuming and required us to create a sense of place that was physically attractive (we planted the orchard, as well as flowers and a community garden, made signs, etc. ... all to make the place look and feel like a shared creative living place), radically inclusive (we made an effort to directly and individually welcome all people, sent emails, handed out flyers, hosted a culturally diverse set of events, etc. and to formalize this into an organizational culture of inclusion), engaging (we planned diverse events at which we had food, cooking, music and dancing, art, gardening, etc.), and open to outsiders' input (like the Community Meal, as well as other community-initiated events). But such an ongoing output of energy could not be sustained by volunteers, and as is common with non-profits (Kanter and Sherman 2016) and activists (Cox 2011), we became burned-out. As for our paid staff, they had to focus on income generation (either through the commercial kitchen members or classes), not community outreach. And so, despite our initial successes, I would evaluate CLiCK as having over the last few years lost momentum in this area. However, as mentioned, CLiCK's new Executive Director (ED) shares our vision of radical inclusion and is working her way back to the community focus that she called "CLiCK's social justice roots", which were what prompted her to join our board back in the very beginning (she had to step down due to the demands of her full-time job at the shelter).

As mentioned in the previous chapter, our community outreach also became plagued by conflicts with some of our community partners, which is also not un-common among non-profits (Plyler 2006). I am putting these conflicts here under 'failures' and linking them to Gramsci's 'time of monsters' because I want them to be seen not as the fault of specific individuals, including Tina and me, but again the result of larger social issues. I understand these monsters to be the result of not being fully part of the dying old world, but also not yet being rooted in the new, thereby inhabiting the spaces of Keating's (2012) 'thresholds'. Additionally, I see many of the conflicts we did have with other non-profits, and more specifically, with other white women working for them, as extensions of the different ways in which we all navigated our roles 'as white saviors'. In recognizing the historical roots of whites acting as saviors under colonialism and continuing up to today, United States–based Nigerian author Teju Cole stated in *The Atlantic* that "...a nobody from America or Europe can go to Africa and become a godlike savior, or at the very least, have his or her emotional needs satisfied" (Cole 2012, para. 11). The same can be said for much of the social service work that is done domes-tically in that, as discussed, the model is of whites, in particular white women, saving people of color and thereby having their emotional needs met by fulfilling their prescribed intersecting gender, racial, and middle-class roles. Naming it the "White Savior Industrial Complex", Cole recognizes that on the global stage this phenomenon acts as a "...valve for releasing the unbearable pressure that builds in a system built on pillage", while never intending to address the actual structural inequalities. Likewise, Cole states, "If we are going to interfere in the lives of others, a little due diligence is a minimum requirement", including having the "...awareness of what else is involved" (para. 19). Yes, Tina and I did spearhead creating a non-profit to serve our collective community and can be seen as performing the roles of 'white women saviors'; however, just as when actors who are assigned roles can choose how they perform/embody them, so too did we attempt to bring 'awareness of what else is involved' and thereby we worked to challenge how we embodied our intersecting socially scripted roles.

In my own case, as a progressive who, as mentioned, can be the most resistant to looking at their own racism, I have worked hard to unpack my racist socialization and to do what Rhonda Magee (2019) refers to as the "inner work of racial justice" in her book of the same title. Magee recognizes that, "We need help developing the capacity to be able to listen to the very different stories of others with compassion; to have conversations across lines of real and perceived difference that help and heal..." (p. 21). In this manner, Tina and I have, as discussed, made it a priority to practice radical inclusion and to have 'conversations across lines of real and perceived difference that help and heal' and to also look at the bigger structural picture. As such, we have sought to see our roles not as Danaher et al. (2007) describe them, as in being 'heroes of the story', but rather as a *part* of a much larger collective story in which we were merely attempting to do our part *with* others, not for them. That said, this solidarity approach is not adopted by all liberal/progressive whites, as became quite apparent to us, not even all those on CLiCK's board or other nearby non-profits. As such CLiCK, hence us whites involved with it, has fallen short of "white people" actively "uprooting racism" as outlined in detail by Leah Penniman in her seminal book *Farming While Black: Soul Fire Farm's Practical Guide to Liberation on the Land* (2018). However, recognizing this is not an excuse for not taking more proactive steps to do better, especially after what National Public Radio (NPR) has referred to as the 'Summer of Racial Awakening' (Chang et al. 2020). In addressing this failure, CLiCK's staff, a board member (a white male), and myself (two are white, one is Latina) are currently offering a Zoom class to other area organizations/non-profits (whose staff and board members are mostly white) that do food-related work titled 'Racism in the Food System' loosely based on Penniman's work (2018) and Soul Fire Farm's approach,[2] along with other materials.[3] Since this is the first time CLiCK is "actively" working to "...resist white supremacy" (Penniman 2018, ch. 16), both within ourselves as whites and within the organization itself, I think it is fair to say that CLiCK has yet to fully move beyond 'white fragility' to "allyship" (Penniman 2018, ch. 16). Thus, I would rank this as a failure, one that currently places aspects of CLiCK more in line with the Alternative Food Movement (AFM), as opposed to the Food Justice Movement (FJM).

Recognizing that we did not specifically engage in anti-racist training for our board or staff, in practicing 'radical inclusion' and prioritizing 'power-*with*' others rather than 'power-*over*', I still believe, as mentioned above, that we did challenge the more typical non-profit practices, contributing to our conflicts with other local community organizations and other middle-class white women who worked at them. We sought, as our current Executive Director (ED) once said, to "Think more like a community than as individual non-profits", and yet the old model of individuals and/or individual organizations being competitive, controlling, self-serving, and non-collaborative made our desire to build trust and work collaboratively with them very difficult. In fact, the nature of our conflicts was that we felt that we had been 'used and abused' by some other non-profits, making us no longer want to 'play', even though we shared similar objectives. For example, in planning collective events involving a number of other non-profits, we consistently felt that we would be left doing most of the work, while also not getting our share of the rewards. What in planning was seemingly inclusive, democratic, and fair would become the opposite in actual practice, and yet when we tried to address this disconnect, there would be a lack of acknowledgment and outright defensive denial. In looking back at these conflicts now, I recognize that they can be understood as being connected to the white fragility–savior complex in that they manifested as others' desires to maintain power and control over supposed shared/collaborative events. In fact, as briefly mentioned in the previous chapter, it can be argued that for many whites, and middle-class white women in particular, the need to be 'saviors' to those deemed less fortunate than themselves in terms of the intersections of race, class, and gender can help to counter their feelings of 'fragility'. As such, by connecting aspects of these conflicts back to varying manifestations of the white fragility–savior complex, in particular the tendency of white women when confronted with racism or other examples of acting in ways that are the opposite of what they claim to declare their innocence and to even shed "…white women tears" (DiAngelo 2018, p. 131), it further clarifies the nature of our conflicts. For even though these conflicts occurred *between* white middle-class women all working in a non-profit/progressive practices capacity, the defining influences of white racism, as well as the ways patriarchy (as well as social class), unequally and differently shaped our relationships to power as white women, were ever present. Recognizing these intersections of racism, sexism, and economic inequalities is not something that DiAngelo does, as previously mentioned, however I do agree with her that addressing white racism, hence fragility, along with all the other forms of oppression " …is a messy, lifelong process" (p. 154). And yet if we are serious about doing social justice work, unpacking both our own racisms and those institutional expressions within ourselves, our non-profits, as well as our larger society, then we must "…align [our] professed values with [our] real actions" (p. 154).

On this theme, CLiCK needs to continue self-evaluating its staff and its board members from an intersectional perspective in order to ameliorate the continuing lack of racial diversity. This issue has been discussed previously, but it remains an ongoing failure for CLiCK and for most non-profits; according to a 2017 report by BoardSource, People of Color (POC) make up a mere 16% of board members, a number that has not changed since 1994 (Biemesderfer 2017). Poignantly, as Vu Le states, "If your board is not representative of the community you claim to serve, then you are furthering the injustice you seek to fight" (Le 2017a, para. 4). I agree with this observation, but as with the earlier claim that non-profits don't care about racial equality, *caring* and *agreeing* don't always translate directly into the results we seek – still, more can and should be done. And it merits stating here, as Le does, that there are other forms of diversity, including "disability, LG-BTQ, age, gender and other identities", as well as social class, education, political affiliation, and professions; in these areas, CLiCK's board has achieved a high level of diversity. Still, more work remains to be done that prioritizes relationship building, authenticity, and intentional outreach (Biemesderfer 2017) across racial/cultural divides.

Another one of CLiCK's failures that still surprises me has been in the area of culinary job training, exemplified by models such as D.C. Central Kitchen, Homeboy-Homegirl Industries/Kitchen, and Hot Bread Kitchen (see Appendix for more details). Right from the beginning, we articulated our goal to have culinary job training, and although CLiCK, as well as several member businesses, has at times hired formerly incarcerated individuals and/or individuals in recovery, a full-fledged culinary training program has not yet been developed. We have also entertained the idea of running an in-house catering company (the former GM, Tina, and I even came up with the name *The 13th Plate*, based again on the pulling of tarot cards) that would hire individuals from the culinary training program, but this idea too still has not manifested. For guidance, we looked to a nearby non-profit in Hartford, Connecticut, called Hands on Hartford,[4] which has a shared-use kitchen, a public café, and a catering company, as well as a food pantry and other social service resources. The former GM and I once went to visit to create a relationship and learn about how they started their café/catering company, only to find out that as a long-time non-profit (started in 1969) they had owned a number of properties which they sold, and so, they had over $200,000 in seed money. At this point, we knew why we had failed thus far in starting our culinary job training program/catering company – not through a lack of visioning or effort on our part, but a simple financial truth, as we have never had that level of start-up capital. In fact, overall, we found our cooperative model, in terms of being a non-profit shared-use kitchen and culinary business incubator, has proven challenging for potential financial supporters and even for some members, as previously discussed. As a society, we are more used to the charity, dependency, and *power-over* model; being invited instead to have *power-with* others, to support people's economic endeavors and potential for growth, rather than their immediate physical needs, challenges our social norms.

Additionally, as previously mentioned, from the beginning, we had visions of being able to help our local farmers add value to their produce by acting as a co-packing/small-scale processing facility that would help solve the double-edged sword of food industry labor, as in on the one hand the cost of labor (as opposed to using mechanization) and on the other hand the food industry's 'excessive exploitation', through job training. In other words, those doing the processing would also be part of some state or federally funded job training program that would both offer a decent wage and the potential for future employment in the food industry. I still think the time for this idea will come, especially since the restaurant industry has in 2020 lost over 5 million jobs due to the COVID-19 pandemic (Zhang 2020), and some of these people may seek new culinary skills and opportunities, as well as new ways of engaging with the food industry, including being involved with similar shared-use cooperative kitchen models such as CLiCK. However, this is not something CLiCK can do alone, nor can it be achieved in any significant ways for as Holt-Gimenez (2017) states,

> Unless we change the underlying value relations of our food system – the contradiction between food as essential for human life and food as a commodity – we will be working on the margins of a system that is structurally designed for profit rather than need, speculation rather than equity, and extraction rather than resilience.
>
> (p. 69)

Small-scale cooperative kitchen models like CLiCK and others (see Agyeman and Loh 2017) are a start at such change, but the challenge is to upscale them within the existing dominant capitalist market that tends to taint all that it encompasses, even as it also makes products more accessible, hence more equitable. Hence, the contradictions continue, as in the case of organic farming, which has become both increasingly industrialized, accessible, and profitable, while still being distinct in terms of the actual farming *practices* from non-organic (Guthman 2014).

Linked to this failure has been our inability to deliver ongoing nutritional education (more than CLiCK's current single classes or periodic programming) and to do so with larger institutional support from University of Connecticut (UCONN), either in a financial way or in a research-based institutional way. UCONN Extension and Eastern Connecticut State University, in particular their Community Outreach Program, have both been supportive members of CLiCK, both as institutions and through the volunteered time and board or committee involvement of several of their staff members. Faculty from both universities have served on CLiCK's board, but still more could be achieved in terms of internships, educational programs, and research agendas that would ideally benefit all parties. I have been very involved over the last five years with UCONN's service-learning program and have worked to make CLiCK an ongoing community partner for service-learning projects, but with COVID-19, the future of such initiatives remains unknown even as the commitment to such partnerships remains. In addition, sponsoring interns and service-learning projects is challenging, requires the investment of staff time, and can often turn into what is known as the "hit it and quit it" (Cushman 2002) model, in which researchers/students use community partners for their own objectives. Given my connection with CLiCK, I have been able to somewhat ensure that this has not been the case, for the most part, but not all the internships or service-learning projects at CLiCK have originated through me. Furthermore, as Brenda Bushouse (2005) has found in her research on community partners' experiences of service-learning projects, in order for community partners to fully benefit, "...time and resources" must be invested to ensure "... ever more complex relationships" (p. 32), particularly finding ways to cover the cost of "...staff time in funding strategies" (p. 40). In short, universities such as UCONN need to find more direct ways to financially support their 'community partners' and not just use them as a means to their students' ends. And so once again, CLiCK's failure to gain greater institutional and financial support from its local universities reflects a broader failure that is socially endemic.

> *I believe the next step that CLiCK needs to do is to expand our presence in the community. We need to let our community brothers and sisters know that we are there for them, that we are there to help them achieve their goals and take off as independent and sustainable business people regardless of their background, native language, race, or gender.*
> Middle-class Latino board member
>
> *Reaching deeper into the low-income community and serving folks from a host of different cultures would be rewarding and tremendously valuable. But I fear the cost of doing business (paying staff, electric and gas bills, etc.) seems to preclude offering space and education to folks for free or at a significantly reduced rate. Serving that most underserved population could encourage inclusion and increased stabilization of immigrant and refugee communities – they likely have little exposure to our regulatory requirements.... But CLiCK needs to find ways to increase revenues, and possibly determine the degree to which it can be self-sustaining (or not) before it can undertake a non-revenue-producing program.*
> Middle-class white female volunteer and legal advisor

Old models, monsters, and ideas whose time has not yet come from without to within

We also had ideas for composting food waste through a cooperative business that would provide jobs for the homeless, in collaboration with our now-ED who at the time was ED at the No Freeze Shelter. Soon after buying the building, we accepted a donation of an Earth Tub composter, made by Green Mountain Technologies (the Earth Tub has since been discontinued, gone from success to failure), from Eastern Connecticut State University's Institute for Sustainable Energy. An Earth Tub requires three-phase electric power to run its fan, but our building had two-phase power, a technical obstacle that we never overcame despite many efforts on my part and our former GM's (including my attempt to get faculty at UCONN working in biochar and anaerobic digesters to adopt the project and make it happen, with no luck). All of CLiCK's members' food scraps therefore sadly go into the standard dumpster, as our riverside location (and the risk of food runoff) and the potential for vermin make open composting untenable. There is a small hand-rotating composter by CLiCK's community garden, but it is too small to be used commercially. Again, CLiCK's troubles echo broader patterns; according to a 2019 article in the *Hartford Courant*, Connecticut throws away 520,000 tons of food every year (an increase from 321,500 tons in 2010). There are companies trying to collect food waste and compost it, like *Blue Earth Composting* in Hartford (Hladky 2019), but such changes are very slow. Still, nationally the tide is turning, as the Environmental Protection Agency reported that "...composting of food rose from 1.84 million tons in 2013 (5.0% of food) to 2.6 million tons (6.3% of food) in 2017" (EPA), and so eventually, I hope that this failure of CLiCK's will become a success along with increased composting around the nation.

Composting, or "in-sourcing fertility" as Joel Salatin of the highly successful Polyface Farm describes it, is one of the 10 markers of what he refers to as the "integrity food system" as it is essential to "...closing the carbon leak" (Salatin 2012, para 26), but perhaps sustainability's most recognizable image and assumed green energy 'solution' is that of solar panels. As such, CLiCK sought to install solar panels on its roof, working with a non-profit called RE-Volv,[5] which works with 'student ambassadors' who do crowdfunding to help offset installation costs. Once the costs of the panels and installation are paid off, RE-Volv uses those funds to help offset the next project, creating "a revolving fund for social energy that continually perpetuates itself building more and more solar". The project and the student ambassadors were sponsored by a UCONN faculty member, and we had many phone meetings with RE-Volv, but in the end, the board voted not to go ahead due to critiques of the current battery technology, fears of unforeseen costs, and CLiCK's overall precarious financial standing. I voted to go ahead with the project, even though in my 'Sustainable Societies' course I use Ozzie Zehner's argument in *Green Illusions: The Dirty Secrets of Clean Energy* (2016). In this highly provocative and well researched book, Zehner proposes that we are not asking the right questions about 'alternative energy' and, moreover, that a focus on solar panels overlooks social issues such as energy and resource consumption, as well as social inequality and in particular gender inequality. To illustrate, in the conclusion of his chapter 'Solar Cells and other Fairy Tales', he states, "Solar cells shine brightly within the idealism of textbooks and the glossy pages of environmental magazines, but real-world experiences reveal a scattered collection of side effects and limitations that rarely mature into attractive realities" (p. 30). In short, "... there are many routes to a more durable, just and prosperous energy system, but the glitzy path carved out by today's archaic solar cells doesn't appear to be one of them" (p. 30). This point is of course highly controversial, but it is important to recognize, as Zehner points out, that solar energy is renewable but solar panels are not and are in fact highly toxic (Shellenberger 2018). However, despite knowing this, I voted to work with RE-Volv for exactly the reasons Zehner (2016) points out, as in my own attachment to the trope of sustainability and for the sake of the public's *perception* of CLiCK. This perception would have spoken to our society's desires for quick and easy "...technological interventions instead of addressing the underlying conditions from which our energy crises arise" (p. xvi), as opposed to the hard and slow work of actual JS, beginning as we were trying to do with the JS imaginary. There is still the possibility that CLiCK will secure solar panels to reduce its own electrical costs and enhance its sustainability image, but based on Zehner's critique, such an act would ultimately remain symbolic.

> *To be even more "sustainable", CLiCK might seek to create and store energy (likely through PV and batteries; perhaps hydro solar augmenting) so that the site can function during extended breakdowns in power production or distribution.*
> Middle-class white male volunteer

A final failure that stands out at this challenging social and economic time has been the failure to secure significant funding to ensure our future sustainability. When considering this, I am reminded of an event for non-profits organized by a bank foundation that I attended a few years ago, at which Vu Le (whose work I have already mentioned) had been the keynote speaker. He suggested getting to know the directors of foundations better, and so I took the opportunity after the event to invite the director of the bank foundation to come visit CLiCK. This director, a white female, had never been to CLiCK, although her foundation had in the past given us a number of small grants, amounting to about $10,000 in total. In discussing CLiCK's financial future, she said we needed to find "legacy donors", noting that baby-boomers are aging and looking for places to donate their funds. In short, she proposed we find some local rich people who might be willing to make substantial donations to assure our financial future. Although I thought this sounded encouraging, as always there is the proverbial shadow between the idea and the reality.

Interestingly – and ironically, given our mission (or not in that it speaks to another form of radical inclusion) – CLiCK has had connections with two wealthy baby-boomer white males, neither of whom have become 'legacy donors'. One used our kitchen to start a chocolate business that became successful and eventually moved to its own location. Before moving on, I had learned of this member's financial status when one of the small local foundations contacted me to ask if I knew of his past role as a very successful financial planner. I told her truthfully that I had no idea who he was (later, of course, I looked him up) but that we treat all members the same, meaning that if he sought to financially support us beyond the standard membership and usage fees, the idea would have to come from him. It never did, and in fact, CLiCK is not even mentioned on his business' webpage as part of his success story, which almost feels more of a loss than the lack of a legacy donation.

Another connection with a potential 'legacy donor' was in relation to a kitchen member, whose business partner owns a local farm and multiple businesses around the world. Since she worked closely with him (and sourced her chickens and local vegetables from his farm), we set up a meeting to ask if he would be interested in working with us on a project to connect to town water and sewage (CLiCK is currently applying for an Economic Development Administration Grant for this purpose, which unfortunately requires a partial matching of funds, making it a stumbling block). When presented with the prospect, he offered instead to buy the building in order to knock it down to make his own kitchen out of which his kitchen member partner would work, suggesting that CLiCK could find somewhere else to operate. At a loss for what to say, we all awkwardly laughed as if he could not have been serious (he was) and then explained based on our non-profit status, our cooperative values, and our social justice commitments why that would not work. After that, he seemed to indicate facially that he was done with the conversation and so before it officially ended, I asked if he would consider making any level of donation, to which he simply replied, "No, not at this time".

I share these two stories not to vilify the people involved, nor to support what Benjamin Priday, an economist who studies charitable giving, sees as the commonly held belief that the rich are not as generous as the rest of us. Priday's analysis of the data actually indicates that based on "…the share of income they give away, …the rich are at least as generous as the poor" (Priday 2020, para. 28). Rather, I share these interactions to illustrate the disconnect between what 'funders' think and the reality on the ground. Here, I look once again to Le, who in his blog "Let's Stop Recognizing Donors by Donation Levels" (Le 2019; also see Le 2017b), argues that, "…the way we do things communicates our values…" and when we prioritize and privilege the highest donors (or those we think could be or should be), we send messages such as "…money is the most important contribution anyone can make" (Le 2019, para. 11) or "…those who give more money deserve special praise, acknowledgement and treatment" (Le 2019, para. 6). However, as he goes on to explain, we cannot know everyone's circumstances and the value and meaningfulness of people's donations are not absolute but relative. As such, having "…more 'smaller' donors is an indication of broad community support" (Le 2019, para. 14) (like Bernie Sanders), as opposed to going after one or two very wealthy individuals. In yet another timely blog written earlier this year, he states, "It's time we fundraise in a way that doesn't uphold white moderation and white supremacy" (Le 2020). Le goes even further stating, and I quote in full:

> By centering the comfort of donors, most of whom are white, we *perpetuate white saviorism, poverty tourism, and inequity* [italics in original] while allowing our donors to avoid confronting difficult realizations like the fact that wealth is built on colonization, slavery, and other forms of injustice. In order for our sector to achieve its mission of creating the world we want, we must ground our fundraising practices in equity, anti-racism, and racial and economic justice.
>
> (para. 3)

With this lens, although both potential 'legacy donors' benefited from CLiCK's low prices, community/business support (including helping the chocolate businesses get started), and our cooperative values, we can also take it as a compliment that they did not choose to support us further, indicating that we stand in opposition to their 'old model'. In sum, although we have in one sense failed in this area of 'legacy donors', had we succeeded in having these two individuals give us money, it might have been emblematic of Dylan's seductive line from his song *Love Minus Zero/No Limit*: "There's no success like failure, and that failure's no success at all" (Dylan 1965). For in teasing out, as I have tried to do in these two last chapters, CLiCK's threshold and entangled 'successes, failures, and unknowns (thus far...)', it becomes ever more evident that the answers to all my questions will in many ways be 'no answers at all', as will be *inconclusively* explored in the Conclusion.

After listening to their ideas of creating a licensed kitchen for food entrepreneurs, farmers, and food educators; I thought it was brilliant! I became the Treasurer of CLiCK for the first 2 &1/2 years. Because CLiCK was a new type of business, a non-profit commercial kitchen, there were many difficulties securing finances, paying taxes, etc. It was really challenging, more than I was equipped to handle. So, I left the board.

Being on the CLiCK board from the very beginning was pretty special. I feel proud to have albeit a very small part of the inception of CLiCK.

Middle-class white female former board member

CLiCK might be great at providing local schools with materials about our local farms and our local agricultural economy. Also, providing food to our large universities from our local farms would help build our local economy.

Middle-class white female politician and supporter

Notes

1 https://www.goodreads.com/author/quotes/15291385.Hafez
2 https://www.soulfirefarm.org/food-sovereignty-education/undoing-racism-farmers-immersion/
3 https://foodfirst.org/wp-content/uploads/2016/03/DR1Final.pdf
4 http://www.handsonhartford.org/
5 See https://re-volv.org/

References

Agyeman, J. and Loh, P. 2017. 'Urban food sharing and emerging Boston food economy', *Geoforum*, 99: 213–222.

Bailey, J. and Flowers, L. 2016. 'The people's supper' [online]. Available at: https://thepeoplessupper.org/new-home-page [Accessed September 30, 2020]

Biemesderfer, D. 2017. 'The Nonprofit Sector's board diversity problem' [online]. *The Center for Effective Philanthropy*. Available at: https://cep.org/nonprofit-sectors-board-diversity-problem/#:~:text=A%20recent%20BoardSource%20report%20showed, nonprofit%20board%20members%2C%20nearly%20identical [Accessed September 30, 2020]

Bushouse, B. 2005. 'Community nonprofit organizations and service-learning: Resource constraints to building partnerships with universities,' *Michigan Journal of Community Service Learning* 12(1): 32–40.

Chang, A., Martin, R. and Marrapodi, E. 2020. 'Summer of racial awakening' [online]. NPR Available at https://www.npr.org/2020/08/16/902179773/summer-of-racial-reckoning-the-match-lit [Accessed October 30, 2020]

Cole, T. 2012. 'The white savior industrial complex'. [online] *The Atlantic*, March 21. Available at: https://www.theatlantic.com/international/archive/2012/03/the-white-savior-industrial-complex/254843/? [Accessed September 30, 2020]

Cox, L. 2011. *How do we keep going? Activist burnout and sustainability in social movements.* Helsinki: Into-ebooks.

Cushman, E. 2002. 'Sustainable service-learning programs', *College Composition and Communication* 54(1): 40–65.

Danaher, K., Briggs, S. and Mark, J. 2007. *Building the green economy: Success stories from the grassroots.* Sausalito, CA: PoliPointPress.

DiAngelo, R. 2018. *White fragility: Why it is so hard for white people to talk about racism.* New York: Beacon Books.

DuPuis, M., Harrison, J. L. and Goodman, D. 2011. '"Just food?"'. In Alkon, A. H. and Agyeman, J., eds. *Cultivating food justice: Race, class, and sustainability.* Cambridge, MA: MIT Press, pp. 47–64.

Duma, A. 1997. *The three musketeers.* Knoxville, TN: Wordsworth Editions.

Dylan, B. 1965. 'Love minus zero / no limit', *Bringing it all back home.* New Your, NY: Columbia Records.

Godfrey, P. 2017. 'Radical pedagogical homesteading: Returning the 'species' to our 'being''. In Haltinner, K. and Hormel, L., eds. *Teaching economic inequality and capitalism in contemporary America.* New York: Springer, pp. 91–103.

Guthman, J. 2014. *Agrarian dreams: The paradox of organic faming in California.* Berkley: University of California Press.

Hladky, G. 2019. 'Connecticut throws away 520,000 tons of food every year. Hartford-based Blue Earth Compost hopes to change that'. *Hartford Courant*, March 28.

Holt-Gimenez, E. 2017. *The foodies guide to capitalism: Understanding the political economy of what we eat.* New York: Monthly Review Press.

Kanter, B. and Sherman, A. 2016. *The happy healthy nonprofit: Strategies for impact without burnout.* New York: Wiley.

Keating, A. L. (2012). *Transformation now!: Towards a post-oppositional politics of change.* Ann Arbor, MI: University of Illinois Press.

Le, V. 2017a. '7 things you can do to improve the sad pathetic state of board diversity' [online]. *Nonprofit.af.* Available at: https://nonprofitquarterly.org/7-things-can-improve-sad-pathetic-state-board-diversity/ [Accessed September 30, 2020]

Le, V. 2017b. 'How donor-centrism perpetuates inequality, and why we must move toward community-centric fundraising' [online]. *Nonprofit.af.* Available at: https://nonprofitaf.com/2017/05/how-donor-centrism-perpetuates-inequity-and-why-we-must-move-toward-community-centric-fundraising/ [Accessed September 30, 2020]

Le, V. 2019. 'Let's stop recognizing donors by donation levels' [online]. *Nonprofit.af.* Available at: https://nonprofitaf.com/2019/10/lets-stop-recognizing-donors-by-donation-levels/ [Accessed September 30, 2020]

Le, V. 2020. 'It's time we fundraise in a way that doesn't uphold white moderation and white supremacy' [online]. *Nonprofit.af.* Available at: https://nonprofitaf.com/2020/06/its-time-we-fundraise-in-a-way-that-doesnt-uphold-white-moderation-and-white-supremacy/ [Accessed September 30, 2020]

Magee, R. 2019. *The inner work of racial justice: Healing ourselves and transforming our communities through mindfulness.* New York: TarcherPerigee.

Penniman, L. 2018. *Farming while Black: Soul Fire's Farm's practical guide to liberation on the land*. VT: Chelsea Green Publishing. Available at: https://ebookcentral.proquest.com/ [Accessed September 30, 2020]

Plyler, J. (2006). 'How to keep on keeping on: Sustaining ourselves in community organizing and social justice struggles', *Upping the Anti*, 3: 123–134. Available at: https://uppingtheanti.org/journal/article/03-how-to-keep-on-keeping-on/ [Accessed 30 September 2020].

Prida, B. 2020. 'Economics professor for the conversation' [online]. Available at: https:// liberalarts.tamu.edu/blog/2020/05/20/rich-folks-arent-that-stingy-after-all/ [Accessed September 30, 2020]

Salatin, J. 2012. Farming success: Joel Salatin's top 10 markers. Available at: https://www. ecofarmingdaily.com/farm-management/business-planning/joel-salatin-10-farming-success/ [Accessed 30 September 2020].

Shellenberger, M. 2018. 'If solar panels are so clean, why do they produce so much toxic waste?' [online] Forbes. Available at: https://www.forbes.com/sites/michaelshellenberger/2018/05/23/ if-solar-panels-are-so-clean-why-do-they-produce-so-much-toxic-waste/?sh=3ea4f57b121c [Accessed September 30, 2020]

Zehner, O. 2016.*Green illusions: The dirty secrets of clean energy*. Lincoln: University of Nebraska Press.

Zhang, J. 2020. 'The restaurant industry Lost 5.5 million jobs in April: One in four jobs lost across the U.S. last month were in restaurants and bars'. *Eater.com*, May 8th. Available at: https://www.eater.com/2020/5/8/21251960/restaurant-industry-jobs-lost-unemployment-april-coronavirus-pandemic [Accessed September 30, 2020]

Conclusion

Interconnections now and beyond…

> "Ever not quite" has to be said of the best attempts made anywhere in the universe at attaining all-inclusiveness.
>
> William James

> Not all of us can do great things. But we can do small things with great love.
>
> Mother Teresa

Interconnections and the [im]possibilities of just sustainabilities[1]

As I write this conclusion in the fall of 2020, I agree with the recent cuttingly pointed observation by Fintan O'Toole of the *Irish Times* that there exists a new emotional response around the world for the United States and that is "pity" (O'Toole 2020, papa. 1). The West coast is facing unprecedented fires due to the continually unspoken issue of climate change and overdevelopment, while the now-widespread ability to film police killings of African-Americans is bringing the ongoing legacy of structural racism to new levels of public exposure. Additionally, COVID-19 continues to spread both nationally and globally, even as it is in many ways still denied (much like climate change) and the full social and economic repercussions downplayed. Consequently, the social fabric of this nation seems to be ripping at the seams spurred on by the increasingly grotesque and extremist behavior of the current Trump administration (hopefully soon ending). Additionally, the industrial food system increasingly wreaks havoc on the Earth's living systems of soil, water, air, climate, and of course all the dependent biodiversity, including our own bodies, thereby further solidifying systemic food injustices and inequalities. Such global and local crises can feel overwhelming, paralyzing, and ultimately heartbreaking, and yet I believe that engaging in just sustainabilities (JS), intersectionality, and their evolving "interconnectedness" (Keating 2012) as theoretical guides for continual reflexive *praxis* can guide us to live into the solutions. Thus, for me and many others, the act of working together on and *with* CLiCK has been our catharsis, our "constructive" "resistance", our "collective agency" (White 2018, pp. 6–7), our way of transmuting our distress into the *possibility* for creating the alchemists' gold. From a 'good idea' collectively envisioned by the board of the Willimantic Co-op over ten years ago, to years of Tina and I and a few others trying to find ways to make it happen, to forming a board and creating a non-profit, to buying a building, to securing grant funds and hosting extensive community events over the years, to helping incubate a significant number of culinary businesses and running a wide variety of nutrition and cooking classes, to planting and growing an orchard and community garden, to creating a labyrinth, a mural, and more, to most importantly engaging in 'radical inclusion' and formally creating a 'culture of inclusion', CLiCK has *become* and continues to be Real.

Case in point, this week Tina and I are silk screening CLiCK's logo onto 50 aprons as part of a 'virtual fundraiser', and I will attend a grants committee meeting over Zoom. I am also, as mentioned, currently offering along with several others (two young female staff members – one white, one Latina – and an older white male board member) a class on 'Racism in the Food System' and working with a Nicaraguan artist who is at University of Connecticut (UCONN) under the Artist's Protection Fund to do a mural on farm worker liberation for the front of CLiCK's building. And so, the story goes on and hopefully, on. Remembering the concepts from Barad (2007) in relation to beginnings, I now also apply them to endings – this may be the end of this book, but CLiCK as the ongoing creation of many individual ideas made collectively manifest will not end, even if the actual organization does.

The reason for this is that people will *always* need to "grow, cook, and share" food, community, and ultimately stories. Additionally, unlike the more individualized, hence more self-serving, approach of the Alternative Food Movement (AFM), CLiCK, along with many *many* other expressions of the Food Justice Movement (FJM; although as identified in relation to 'uprooting racism', CLiCK still has much work to be done), is a collective enterprise, bringing disparate people and communities together to find mutually beneficial ways of solving our most insidious and inseparable societal and environmental problems, of which food is central. As the currency of our physical and social beings, to repeat, food is both the "problem and the solution" (Finley 2013). Food is the stuff from which we collectively create ourselves and our stories and therefore it matters whether or not it is destructive and/or beneficial or both to ourselves, each other, and the world. As Harari (2011) has identified, it is our nature as singularly *social beings* (although I would add that many other species are social but not to our storytelling level) that has resulted in us being the dominant species of the world; yet, it is also this ability that is challenging our survival, as our 'imagined' social and cultural worlds (that includes the constructions of all our inequalities) have become more 'real' to us than the natural one that physically, emotionally, and spiritually sustains us (this disconnect is particularly strong in the West). For as Abrams (1996) argues, "…we are humans only in contact, and conviviality, with what is not human" (p. ix). Therefore, we must, as he implores, "…renew our acquaintance with the sensuous world" (p. ix), a world that I would add includes food.

Hence, the increasing importance of places and organizations like CLiCK, as *physical* places (Agyeman and Warner 2002) that have the potential to find new and innovative ways to collectively ground us back into our food, our bodies, and our stories. In this manner, Mark Winne (2019), in the conclusion to his most recent book *Food Town, USA: Seven Unlikely Cities that are Changing the Way We Eat*, calls for readers to "Nurture, nest and incubate the creative impulse of the people so that a thousand ideas bloom" and to do so by "…promot[ing] forums to raise up ideas, provide technical assistance, and to offer mutual support to budding entrepreneurs" (p. 188); all of which CLiCK has tried to do and continues to do, albeit on a small scale. The ways we have tried to do this have been by moving collectively from the theories in our minds, into the practices of our bodies. Therefore, as Freake and I wrote back in 2016,

> One of the most fascinating aspects of working on and with CLiCK for both of us has been to watch aspects of our academic disciplines as 'theories' come into being as practices and as members of the board to further guide and revise those theories as new issues emerge and new practices are needed.
>
> (p. 126)

In building on this earlier work, I stated in the Introduction that my goal was to discern whether or not *we* as CLiCK (that is now its own entity) 'have succeeded in creating, experiencing, and sustaining aspects of JS, as defined by its four principles'. If so, I proposed to investigate all the ways and whys of 'yes' and if not all the ways and whys of 'no', including an attempt to discern if 'yes', then what such a state might actually *feel* like. Additionally, I proposed that I would engage in an intersectional analysis in order to further assess our successes, failures, and unknowns (thus far), and as such assess our *praxis*. Then, later on, I posed additional questions more specifically about our creation of CLiCK in order to use them as lines of inquiry, although I did not assert completed answers. And now that I have come to the end of exploring a small fraction of my emic and etic stories, occupying, as Keating (2012) states about her threshold theories, "…ambivalent insider-outside oscillations…." (p. 11), I think it is again clear not only that there are no definitive answers, but as also proposed in the Introduction, I should not be the one to ultimately offer them even if they existed. Rather, I should leave it up to all the possible positionalities involved, as in up to the reader, up to those who answered my survey questions, up to those who have been to CLiCK, up to those who work/volunteer there, up to those who have incubated/are still incubating their businesses there, and up to posterity to decide. For as argued from the beginning, when engaging with intersectionality, and in addition 'threshold theories', it becomes ever more complex and difficult to speak comprehensively and with full authority about anything, while at the same time, as in this case, not letting go of an overall commitment to achieving JS, at least in theory. This, however, I would assert is not ultimately a failing, but rather should be seen as emblematic of attempting to achieve JS in *praxis*, given that as previously quoted,

"Just sustainabilities is a discourse of hope" (Broto and Westman 2017, p. 648), rather than a preconceived destination.

As such, what I will affirm is that the answers to my questions are all partial, 'imperfect', reminiscent of the 'maybe' given by the farmer in the Zen story, or William James' famous remark about trying to attain an all-inclusive theory, that the best we can offer is *'ever not quite'* (James 1912, p. viii). For like all social entities, and indeed even more so in this case, justice and sustainability are fluid, highly complex, and yet ephemeral, requiring ongoing *"charlas culinarias"* (culinary chats) (Abarca 2004, p. 1; also in Godfrey and Torres 2020), so that they may be nurtured and made real through moment by moment reflexive praxis. As such, JS are not endings that we can say we have reached and can consequently stop the struggle, stop the relationship building, stop the conversation; rather they are processes, journeys, and places we have yet to fully inhabit. Nevertheless, based on all my years of working with Tina and others on and with CLiCK, I can attest, as does Henry Ford, that "Working together is success",[2] even with all the difficulties involved. However, in contrast to Ford's corporate for-profit model, such togetherness should be measured by the degree of "...compassionate actions of its members", as Coretta Scott King so astutely observed (Associated Press 2000, para. 2), as opposed to the amount of profit for the few. In addition, such togetherness now needs to embody 'new models' that are, as Paul Mason argues in his book *PostCapitalism: A Guide to Our Future* (2015), "...cooperative, self-managed, nonhierarchical" (p. 287), although perhaps past examples of food justice have always embodied these attributes (White 2018), including many Indigenous food practices (Vernon 2015). Furthermore, in seeking to 'understand just sustainabilities' and its four principles, working together across diverse communities with the larger goals of achieving expressions of equity, justice, and sustainability has ultimately made me feel fully alive, hence *Real*, despite all the stressors and challenges, or perhaps because of them. Such work is hard, ongoing, 'ever not quite', and yet also cumulative and potentially contagious, spreading in small and web-like ways, often unseen, like the mycelium networks termed by Wohlleben (2015) as the 'wood-wide web', enabling trees to constantly communicate, thereby creating resilience.

Highlighting the role of the collective, as articulated within the FJM, flies in the face of most popular American narratives around sustainability, in which the 'individual reigns supreme'. In fact, as Wade Davis recently stated in his cuttingly astute article *The Unraveling of America*, "More than any other county, The United States in the post-war era [has] lionized the individual at the expense of community and family...the sociological equivalent of splitting the atom" (Davis 2020). Excessive individualism, as promulgated in the USA, therefore challenges collective action in times of crisis as in the current global pandemic (Bazzi et al. 2020), which as it spreads, ironically further increases the need to be 'separate', but to do so 'together'. Yet, when it comes to supposedly achieving sustainability, we are told over and over that it is the individual, the lone atomic particle, whose actions will 'save the planet', 'clean up toxins', 'solve poverty', and now 'count our carbon footprint' through an app (Yoder 2020). This emphasis constantly overlooks the difficult work of organized and collective action that is essential to create structural changes aimed at a more just and sustainable society for all life. Annie Leonard explores this concept in her ingenious short videos, particularly *The Story of Change* (2012), as well as in her article *Moving From Individual Change to Societal Change* (2013); in both, she recognizes the distortion in placing blame and solutions on individuals, as opposed to the collective structures that can only be changed "…by working together" (p. 252). As Agyeman and Evans (2003) states, what is needed is "…a small group applying good ideas [to] spark[s] society's collective awareness" (p. 49), much like, as they point out, Ken Keyes Jr.'s (1984) inspiring parable of the 'hundredth monkey', wherein it is proposed that when enough people begin to enact positive changes, there reaches a tipping point for said behavior to become the new normal (Keyes 1984). And certainly, this has been the goal of CLiCK: to be one such spark. In addition, Julian Agyeman and Sydney Giacalone (2020), in their recently co-edited volume *The Immigrant-Food Nexus: Borders, Labor and Identity in North America*, seek to show

> …the importance of humanizing, multiscalar, and comprehensive scholarship that resists the tired binaries that reify so many false narratives and incomplete understandings plaguing current conversations about the 'place' of immigrants within our nation.
>
> (p. 306)

Likewise, in honestly telling *my* story of CLiCK, which is also CLiCK's story (but certainly not the only version), I too have tried to engage in 'humanizing, multiscalar, and comprehensive scholarship' and storytelling, which 'resists the tired binaries' that tend to simplify and/or objectify the difficulties and complexities in relation to social justice work. In so doing, as Agyeman and Giacalone (2020) also state, I hope to "…catalyze more productive and transformative conversations…" (p. 306) about the ways and means to do 'small things *with* great love', as Mother Theresa (2010) is paraphrased as having once said, and to keep on doing them in ways that further bring to life Keating's 'threshold theories'.

Keating's (2017) 'threshold theories' build on the theory of intersectionality but move as she explores, "…beyond…into a metaphysics of interconnectedness…" and from there to "…inter-relationality….", which she says "…offers a radically inclusive approach" (p. 30). She argues that intersectionality, "…as too often practiced in contemporary theory and practice, inadvertently draws on and reinforces the types of status-quo thinking…" (p. 36) based on our socially constructed identities. As a result, scholars "…mark, classify, and divide: they stop with the labels and do not use this labelling process to generate new commonalities. The differences function like walls, not thresholds" (p. 37). Of course, her point is not to "…condemn intersectionality" as a theory, for it has served to expose painful and complex inequalities, but to critique how it has been used and to assert that "We need multiple stories and multiple tools…We need new stories, new tactics, and new visions" (p. 37). In this vein, I hope that what I have done here in terms of applying JS and intersectionality as theoretical lenses to critically examine the overlapping macro and micro inequalities that form the industrial food system, and thereby shape our communities, including of course CLiCK, has also been an exploration of existing connections. These existing connections, like *alchemic* threads, have come 'imperfectly' together to make CLiCK *Real* (at least for now), while also speaking to a larger 'interconnectedness' that extends way beyond me, CLiCK, our communities, to the larger global desires for justice and sustainability (Hawkins 2008), hence for JS.

Keating (2017) further observes that in this work, "We have no maps, no clear-cut plans, no definitive, absolutely correct solutions" (p. 59), a truism that I apply directly to my/our experiences with CLiCK. However, even without 'maps' or 'clear-cut plans' or 'absolutely correct solutions', I would emphatically argue, as Holt-Gimenez (2017) does, that, "...to change the food-system", and society as a whole, not only do we "...need to understand capitalism", but we need to outgrow it into more collective, "cooperative", 'humanizing, multiscalar, and comprehensive' models as others have illustrated (Agyeman and Giacalone 2020; White 2018). Furthermore, we need to do so with "love", which, as Holt-Gimenez goes on to say, we will have to take on "faith" as being essential if we seek to "... change the world" (p. 24). Personally, I take it as *fact* based on my years working with Tina on CLiCK, for had it not been for our *love* (Agape and Eros) for each other, we would have never had the love (Agape) to give to others/CLiCK. As a society, it seems we spend an inordinate amount of time 'looking for love in all the wrong places' and prioritizing Eros, erotic love, rather than Agape, spiritual love, yet Agape love can happen as simply as one *charlas*, one *heart-felt* conversation at a time, as in by taking time. Therefore, I return to Niebuhr's quote, which has over the years guided me both in theory and in praxis, in word and in deed, but the third part most of all: Nothing we do, however virtuous, can be accomplished alone; therefore, we must be saved by love (1952, p. 63). Throughout this whole journey, including the writing of this book, I have found this to be my most powerful *understanding* of the slow, difficult, and highly complex task of collectively achieving aspects of JS from *within* my own heart, manifesting as Agape love and thus destined to remain 'ever not quite'.

In endings come other beginnings

And so, I end by adding a few new edits in brackets to the earlier words written by Freake and myself in 2016, words that remain true, perhaps even more so four years later:

> CLiCK is a social and economic experiment whose success or failure doesn't just depend on our individual efforts but rather on the larger questions as to whether we as a community and as a society are committed to change our local and national food systems to serve the interests of all life as opposed to the corporate interests of a few. It is too soon to know the answer for CLiCK or for the larger society but we feel that there is a better chance of the answer being 'Yes' if more communities follow CLiCK's lead. CLiCK

My experience working on CLiCK was enlightening and inspiring. That's not to say it wasn't without some frustration and difficulty in the process of developing and bringing CLiCK to fruition. What I found inspiring was working with a core of individuals of disparate personalities and backgrounds who had conceived of an idea and through diligence and perseverance brought it to reality. The "core" through its passion and drive inspired others to join in its mission. The "team" changed along the way with new members dedicated to the cause as well as existing members dropping out. I found the process a tremendous learning experience and am glad to have been a part of the "team". Having owned a small business where I was the sole decision maker was an entirely different mindset than my experience with CLiCK. It was an experience that broadened my understanding of working with a team on a common goal and understanding that there can be many ways to achieve that goal.
Middle-class white male former board member

I would love to see CLiCK become a resource for our community in holding discussions about climate change, food apartheids, health, and the connection to social injustices. I would love to see CLiCK reach a youthful audience and expanding the amount of volunteer work they have for the community. I would love to see CLiCK connect their kitchen and education aspects together for a new project that could involve free food for the community. I would love to see CLiCK teach households how to turn the grass in their front yard into thriving gardens. I would love to see CLiCK grow fruit trees around the community for free, healthy food for anyone walking by. CLiCK has the potential to take the heavy theory of sociology and food justice into praxis, the process of practicing what you preach on the muddy ground.
Middle-class Latina UCONN intern

offers a theoretical model based on 'just sustainabilities' and 'just nutrition' that encourages community control over 'the commons,' be that a kitchen, knowledge, land or anything else essential to the health and well-being of a community. Additionally, CLiCK offers a practical model in terms of what these theories can look like at a grassroots level but which allows for variability to suit the specific social, economic and geographic locations. An increasing demand for local food, [and not only locally grown but also locally cooked, processed, and prepared], coupled with greater public awareness of the importance of environmental issues, including climate change and the continuing roles that racism / poverty play [and the need to structurally and collectively address them] in the United States make us [cautiously] optimistic. The time is [still] right for CLiCK and other initiatives that incorporate *understanding* [italics added] of these issues in their mission and make them central to their daily [and ongoing] practices [praxis].

(Godfrey and Freake 2016, pp. 126–127)

Additionally, and more metaphysically, this 'understanding' also translates for me into a line I would often sing while working at CLiCK, a lyric from Leonard Cohen's haunting song *So Long Marianne* (1967), that if "I forget to pray for the angels, …. the angels [would] forget to pray for us". These lyrics helped to remind me not only to be grateful for all that we had been able to accomplish, but to also remember that we only did *some* of the work. So much more of what happened that enabled CLiCK to 'become *Real*' happened for reasons and in ways I (we) could not and still cannot explain, and so to these unknown forces, the term angels and the magic, hence the multifaceted *blessings* they bestow, seems fitting. This however does not mean CLiCK will stay *Real*, but as long as the prayers, the conversations, the stories, and the resulting social justice *actions* keep going and slowly growing, the quest for and creation of a more just and sustainable food system/society will prevail. For although the notion of 'slow justice' is for the most part seen as 'no justice', as in the legal maxim a 'justice delayed is a justice denied', I would nevertheless like to counter. When it comes to, as previously mentioned, M. L King Jr.'s vision that, "…the arc of the moral universe" (1968), as well as I would add, large-scale social changes or even smaller ideological and behavioral shifts here on Earth, justice is never a completed accomplishment. It is in fact a slow process of 'bending', of expanding intentions and collective actions that can be conceived as countering Rob Nixon's (2011) concept of "slow violence". Nixon uses this concept to specifically articulate the gradual yet globally unequal threats of environmental destruction and climate change. As such, the utility of 'slow violence' can also be applied to many other forms of socially accumulative and intersecting oppressions (Davies 2019). Therefore, I affirm that *slow justice* must also be recognized, not necessarily as a 'denied' destination but as a nascent cumulative journey that can only be fully understood from, as previously mentioned, the "longue durée" (the long term). Consequently, the same can be said for JS, and perhaps even more so. Therefore, I return one final time to Niebuhr's quote, recognizing more fully that "Nothing …worth doing can be achieved in our lifetime…" (1952, p. 63), while also not forgetting that in the end, "… all life is a dream, and dreams themselves are only dreams /¡… toda la vida es sueño, y los sueños, sueños son!" (De la Barca 2008). Dreams can be redreamed and the seeming impossibilities of JS can over time become *possible*, if we collectively continue to change our social structures, hence our values/principles and praxis, moment by moment, breath by breath, dream by dream.

¡Ándele!/Onwards!

Figure 10.1 The Author with Giant Carrots Grown in CLiCK's Community Garden!

***Recipe:* Cooking Up Community/Cocinando Arriba Comunidad**

This is our cooperative community kitchen project that has been 13 years in the making. All aspects are a work in progress. It is a labor of love, and we hope yours turns out just the way you envision, or in some as of yet unimagined variation!

   Prep time: 4 years
   Cook time: 9 years and counting...
   Ingredients:

1 Good idea
7 Cooperative principles
50 Board members from diverse backgrounds
50,000 Volunteer hours
10,000 Intern hours

100,000 Emails
100 Community/Fundraising events
$600,000 Grant funds
1 Building (5,600 sq ft)
$200,000 Building renovations
30 Fruit trees and gardens
1 Mural (with one in the works)
15 Part-time employees
1 Executive Director
60 Commercial kitchen members
200 Cooking and nutrition classes
Unlimited love, compassion, and humor
Leave room for the unknown

Start with a 'good idea'. Identify highly motivated individuals and collectively create mission, by-laws, and policies based on seven cooperative principles and suited to your specific cooperative kitchen needs. Commit to social justice in theory and praxis.

Apply to be a 501c3 non-profit. Wait a few months... Successful?

Recruit board members, commit to cooperative principles, social justice, practice radical inclusion, assess community needs, and meet monthly.

Add in all volunteer and intern hours, mix in 100,000 emails to begin the slow and arduous process of becoming *Real*.

Schedule and carry out a variety of community/fundraising events based on radical inclusion to inform your community of your organization's mission and amenities. Be sure to make everyone feel included and to emphasize that as a *cooperative* you seek to address their needs, while inviting their input.

Apply for numerous grants and funding sources to ensure your longevity. Purchase one building! Or build one!

Discuss and design the dream kitchens for your facility, then amend designs to align with budget.

Plant 30 fruit trees and community gardens. Paint a community mural.

Begin with staffing the organization with part-time employees. Hire one executive director with a passion for the mission. Commit to JS and engage in on-going reflexive food justice praxis. Continually recruit and help to fund/support new kitchen members who will be able to follow their food dreams. Offer community cooking and nutrition classes in English and Spanish.

Generously sprinkle with love, compassion, & humor. Stir with just intentions, stay committed to your idea, keep checking on your mission, yet allow for the unknown.

Prepare to receive and watch your project grow.

Enjoy *collectively!/¡Disfruta colectivamente!*

Notes

1 Taken from the title of a forthcoming book Global [Im]-Possibilities: Exploring the Paradoxes of Just Sustainabilities, 2021, London: Bloomsbury Press, edited by Mary Buchanan and Phoebe Godfrey.
2 https://www.goodreads.com/quotes/118854-coming-together-is-the-beginning-keeping-together-is-progress-working

References

Abarca, M. E. 2004. 'Authentic or not, it's original', *Food and Foodways* 12(1): 1–25.

Abrams, D. 1996. *The spell of the sensuous*. New York: Vintage Books.

Agyeman, J. and Warner, K. 2002. 'Putting 'just sustainability' into place: From paradigm to practice', *Policy and Management Review* 2(1): 8–40.

Agyeman, J. and Evans, T. 2003. 'Towards just sustainability in urban communities: Building equity rights with sustainable solutions', The *Annals of the American Academy*, AAPSS, 590: 35–53.

Agyeman, J. and Giacalone, S. 2020. *The immigrant-food Nexus: Borders, labor and identity in North America*. Boston, MA: MIT Press.

Associated Press. 2000. 'King's widow urges acts of compassion'. *Los Angeles Times*. [online]. Available at: https://www.latimes.com/archives/la-xpm-2000-jan-17-mn-54832-story.html#:~:text=%E2%80%9CThe%20greatness%20of%20a%20community, a%20soul%20generated%20by%20love.%E2%80%9D [Accessed September 30, 2020]

Barad, K. (2007) *Meeting the universe halfway: Quantum physics and the entanglement of matter and meaning*. Charlotte, NC: Duke University Press.

Bazzi, S., Fiszbein, M. and Gebresilasse, M. 2020. 'Rugged individualism and collective (in)action during the COVID-19 pandemic' [online]. *The National Bureau of Economic Research*. Working Paper No. 27776. Issued September, 2020. Available at: https://www.nber.org/papers/w27776 [Accessed September 30, 2020]

Broto, V. C. and Westman, L. 2017. 'Just sustainabilities and local action: Evidence from 400 flagship initiatives', *Local Environment* 22(5): 635–650.

Cohen, L. 1967. 'So long, Marianne', *Songs of Leonard Cohen*. New York: Columbia Records.

Davis, W. 2020. 'The unraveling of America' [online]. *Rolling Stone Magazine*, August 6.

Davies, T. 2019. 'Slow violence and toxic geographies: "Out of sight to whom?', Environment and Planning C: Politics and Space [online]. Available at: https://journals.sagepub.com/doi/full/10.1177/2399654419841063 [Accessed September 30, 2020]

De la Barca, P. C. 2008. 'Life's a dream / La vida es sueño'. Edited and translated by Applebaum, S. Mineola, NY: Dover Publications.

Finley, R. 2013. 'A guerilla garden in South Central, L.A' [online] TED Talk. Available at: https://www.ted.com/talks/ron_finley_a_guerrilla_gardener_in_south_central_la?language=en [Accessed September 30, 2020]

Godfrey, P. and Freak, H. 2016. Feeding community: A case study of a shared-use commercial kitchen in Eastern Connecticut. In Bosso, C., ed. *Feeding cities: Improving local food access, sustainability, and resilience*. London: Routledge, pp. 113–128.

Godfrey, P. and Torres, D. 2020. 'Recipes for immigrant lives: Crossing, cultivating, cooking, and culture at a shared-use commercial kitchen'. In Giacalone, S. and Agyeman, J., eds. *The Immigrant-food nexus: Borders, labor, and identity in North America*. Boston, MA: MIT Press, pp. 181–298.

Harari, Y. N. 2011. *Sapiens: A brief history of humankind.* New York: Harper.

Hawkins, P. 2008. *Blessed unrest: How the largest social movement in history is restoring grace, justice, and beauty to the world.* London: Penguin Books.

Holt-Gimenez, E. 2017. *The foodies guide to capitalism: Understanding the political economy of what we eat.* New York: Monthly Review Press.

James, W. 1912. *The will to believe and other essays in popular philosophy* [online]. New York: Longman, Green and Co. Available at: https://www.gutenberg.org/files/26659/26659-h/26659-h. [Accessed September 30, 2020]

Keating, A. 2012. *Transformation now!: Towards a post-oppositional politics of change.* Ann Arbor, MI: University of Illinois Press.

Keyes, K. 1984. *The hundredth monkey.* Coos Bay, OR: Vision Books.

King, M. L. 1968. 'Remaining awake through a great revolution. Speech given at the national Cathedral, March 31. [online]. Available at: https://www.si.edu/spotlight/mlk?page=4&iframe=true [Accessed September 30, 2020]

Leonard, A. 2012. *The story of change.* Available at: https://www.youtube.com/watch?v=DZUN6gQhfvM&ab_channel=FreeRange [Accessed 30 September 2020].

Leonard, A. 2013. Moving from individual change to societal change. *In state of the world 2013: Is sustainability still possible?* The Worldwatch Institute, pp. 244–252.

Mason, P. 2015. *Post Capitalism: A guide to our future.* New York: Farrar, Straus and Giroux.

Neibuhr, R. 1952. *The Irony of American History.* New York: Charles Scribner's Sons.

Nixon, R. 2011. *Slow violence and the environmentalism of the poor.* Cambridge, MA: Harvard University Press.

O'Toole, F. 2020. 'Donald Trump has destroyed the country he promised to make great again'[online]. *The Irish Times,* April 25. Available at: https://www.irishtimes.com/opinion/fintan-o-toole-donald-trump-has-destroyed-the-country-he-promised-to-make-great-again-1.4235928?mode=sample&auth-failed=1&pw-origin=https%3A%2F%2Fwww.irishtimes.com%2Fopinion%2Ffintan-o-toole-donald-trump-has-destroyed-the-country-he-promised-to-make-great-again-1.4235928 [Accessed September 30, 2020]

Teresa, M. 2010. [online]. Available at: https://www.motherteresa.org/08_info/Quotesf.html) [Accessed 30 September 2020].

Vernon, R. 2015. A native perspective: Food is nor than consumption [online]. *Journal of Agriculture, Food Systems and Community Development* 5(4): 135–142. Available at: file:///C:/Users/Owner/Desktop/377-Article%20Text-716-1-10-20160925.pdf [Accessed September 30, 2020]

White, M. 2018. *Freedom farmers: Agriculture resistance and the black freedom movement.* Chapel Hill: The University of North Carolina Press.

Winne, M. 2019. *Food town, USA: Seven unlikely cities that are changing the way we eat.* Washington, DC: Island Press.

Wohlleben, P. 2015. *The hidden lives of trees: What they feel, how they communicate.* New York: Greystone Books.

Yoder, K. 2020. 'Footprint fantasy' [online]. *Grist,* August 26. Available at: https://grist.org/energy/footprint-fantasy/ [Accessed September 30, 2020]

Appendix

These are the questions sent to 50 people via email who have been involved with CLiCK (Commercially Licensed Co-operative Kitchen) over the years in different capacities – 28 replied.

Questions

1   What attracted you to becoming involved (please indicate if you were an employee, a board member, a volunteer, an intern, etc.) with the commercial kitchen idea that eventually became CLiCK? How did you become involved? If you are still involved, say why, and if you have since become uninvolved, say why. (Selected answers in Part I)

2   Do you think CLiCK helps promote social/food justice in the community? If so, how? If not, how not and what could it do to do so? Give examples. (Selected answers in Part I)

3   Do you think CLiCK helps promote sustainability in the community? If so, how? If not, how not and what could it do to do so? Give examples. (Selected answers in Part II)

4   How did/has being involved with CLiCK affected you/your life/your business/your personal growth…etc.? Give examples. (Selected answers in Part II)

5   What is the next step that you would like to see for CLiCK in terms of serving the community? (Selected answers in Part III)

6   Anything else you would like to share? (Selected answers in Part III)

List of other Shared-Use Kitchens that have inspired CLiCK

1   CommonWealth Kitchen
      196 Quincy Street
      Dorchester, MA 02121
      http://www.commonwealthkitchen.org/
2   DC Central Kitchen
      425 2nd St. NW
      Washington, DC 20001

https://dccentralkitchen.org/

3   Franklin County Community Development Corporation
    Western MA Food Processing Center
    324 Wells Street, Greenfield, MA 01301
    https://www.fccdc.org/food-processing-center/

4   Hands on Hartford
    55 Bartholomew, Hartford, CT 06106
    http://www.handsonhartford.org/

5   Homeboys Kitchen/Industries/Foods
    https://shop.homeboyindustries.org/collections/kitchen
    https://homeboyfoods.com/

6   Homegirl Café & Catering
    130 W. Bruno St., Los Angeles, CA 90012
    https://homeboyindustries.org/social-enterprises/cafe/

7   Hot Bread Kitchen
    630 Flushing Ave,
    Suite 210
    Brooklyn, NY 11206
    https://hotbreadkitchen.org/

8   Vermont Food Venture Center
    Center for an Agriculture Economy
    Hardwick, VT 05843
    https://hardwickagriculture.org/farmers-food-businesses/
shared-use-commercial-kitchen

Index

Postscript

At the end of writing this book, an event happened that requires mentioning, as it has further secured my claims as to how much this project has evolved and taken on a life of its own. This year, the CLiCK orchard produced more apples and peaches than ever, and yet, no sooner were we celebrating with harvest plans than something unexpected happened. Trees started to disappear, leaving nothing but chewed stumps. The only explanation – our orchard had been discovered by beavers! They took down and dragged away, apples and all, five trees before texts were sent out with a plan to wire the trunks as a means of protecting them. However, these wire trunk covers were not high enough and so another five trees were stealthily removed. As a result, Tina and I bought more wire and worked until dark one night making the wire as high as we could (about 3–4 feet). Feeling pleased that we had stopped the beavers, we were however disappointed to learn that in the darkness (even though we wore headlamps) we had missed one, which by the following day had been removed. Since then, no more have been taken – perhaps due to our additional wiring or perhaps because the beavers have satiated their winter stores. Yet, despite my sadness at the loss of these 11 trees that I and others had worked so hard to secure, plan, water, tend, and propagate over eight years, like the end of the Velveteen Rabbit, when he finally becomes a living rabbit and is recognized as such by other rabbits, the storming of CLiCK's orchard by beavers brought me an ironic sense of happiness and satisfaction. The reason is I have come to recognize that in the eyes of our beaver neighbors living down the banks of the Natchaug River, CLiCK had unequivocally become *Real*. And for that *interspecies* affirmation, 11 trees were worth the sacrifice.

'P.s. I am happy to add that 15 new fruit trees are going to be this spring (2021), generously supplied by the organization One Tree Planted! (https:// onetreeplanted.org/). We will make sure to sufficiently wire the trunks of these ones-beavers beware!

Index

Note: *Italic* page numbers refer to figures and page numbers followed by "n" denote endnotes.